Aven Nelson

First report on the flora of Wyoming

Aven Nelson

First report on the flora of Wyoming

ISBN/EAN: 9783337268480

Printed in Europe, USA, Canada, Australia, Japan

Cover: Foto ©berggeist007 / pixelio.de

More available books at **www.hansebooks.com**

First Report

...on the...

Flora of Wyoming

UNIVERSITY OF WYOMING.

Agricultural College Department.

WYOMING EXPERIMENT STATION,

LARAMIE, WYOMING.

BULLETIN NO. 28.

MAY, 1896.

First Report on the Flora of Wyoming

BY THE BOTANIST.

Bulletins will be sent free upon request. Address: Director Experiment Station, Laramie, Wyo.

WYOMING
Agricultural Experiment Station.

UNIVERSITY OF WYOMING.

BOARD OF TRUSTEES.

Hon. STEPHEN W. DOWNEY, President, Laramie,	1897
GRACE RAYMOND HEBARD, Secretary, Cheyenne,	1897
OTTO GRAMM, Laramie,	1897
Hon. M. C. BROWN, Laramie,	1897
Prof. JAMES O. CHURCHILL, Cheyenne,	1899
Hon. JAMES A. McAVOY, Lander,	1899
Hon. TIMOTHY F. BURKE, Cheyenne,	1901
Hon. JOHN C. DAVIS, Treasurer, Rawlins.	1901
Hon. CARROLL H. PARMELEE, Buffalo,	1901
State Supt. ESTELLE REEL,	Ex Officio
President ALBINUS A. JOHNSON,	Ex-Officio

AGRICULTURAL COMMITTEE OF THE BOARD OF TRUSTEES.

OTTO GRAMM, Chairman,	LARAMIE
S. W. DOWNEY,	LARAMIE
M. C. BROWN,	LARAMIE

PRESIDENT OF THE UNIVERSITY OF WYOMING.
A. A. JOHNSON, A. M., D. D.

STATION COUNCIL.

A. A. JOHNSON, A. M., D. D.,	DIRECTOR
G. R. HEBARD, A. M., Ph. D.	SECRETARY
B. C. BUFFUM, M. S.,	AGRICULTURIST AND HORTICULTURIST
J. D. CONLEY, A. M., Ph. D.,	PHYSICIST AND METEOROLOGIST
AVEN NELSON, M. S., A. M.,	BOTANIST
E. E. SLOSSON, M. S.,	CHEMIST
W. C. KNIGHT, A. M.,	GEOLOGIST

SUPERINTENDENTS.

JACOB S. MEYER,	LANDER EXPERIMENT FARM
JOHN D. PARKER,	SARATOGA EXPERIMENT FARM
JOHN F. LEWIS,	SHERIDAN EXPERIMENT FARM
A. E. HOYT,	SUNDANCE EXPERIMENT FARM
MARTIN R. JOHNSTON,	WHEATLAND EXPERIMENT FARM
B. C. BUFFUM,	WYOMING UNIVERSITY EXPERIMENT FARM
THE HORTICULTURIST IN CHARGE,	WYOMING UNIVERSITY EXPERIMENT GROUNDS

TABLE OF CONTENTS.

	PAGE.
Introduction	47
Collecting trips	48
Plant zones	58
The plains flora	60
Flora of the foothills	62
The mountain flora	63
The trees of the state	64
The floras of the Atlantic and of the Pacific slopes	65
Introduced plants	66
Hardiness of native plants	67
Floral calendar	67
Botanical work in the state	68
Nomenclature and citations	69
Cryptogams	70
Acknowledgments	70
Explanations	71
List of collecting localities, with approximate altitudes	72
LIST OF SPECIES	75
Appendix to List of Species and corrections	204
New species and varieties	206
Lists of plants reported by other collectors	207
Summary	213
Index to genera	215

ERRATA.

Page 83 read Erysimum for Erisymum.
Page 99 read Glycyrrhiza for Glyceria.
Page 170 read Monolepis for Monolepsis.
Page 199 read Cryptogramme for Chryptogramma.
Page 142 line 5 from bottom read Triassic for Tertiary.

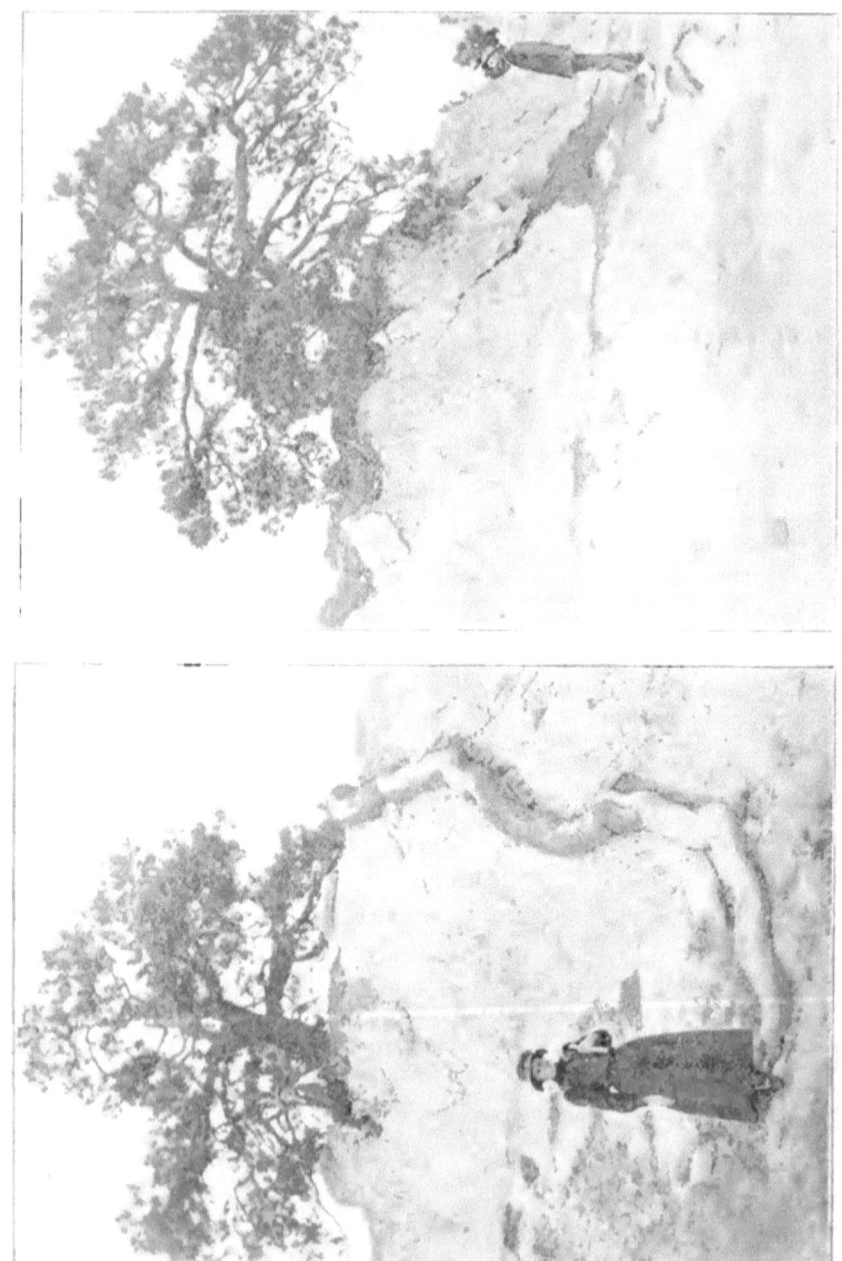

ROCKY MOUNTAIN YELLOW PINE.

Showing remarkable root development. Length from tree to point where root enters the ground, 45 feet. Sand Hills, 12 miles north of Lusk.
From Photograph by Prof. B. C. Buffum.

FIRST REPORT
——ON THE——
FLORA OF WYOMING.

BY AVEN NELSON.

INTRODUCTION.

Among the duties planned by and for the Botanist of the Experiment Station for the year 1894 were the following:

1. To study the Fungi affecting the ordinary farm crops and the best means of combatting the same.
2. To give attention to the weed question, with a view to finding effective methods for exterminating or preventing the spread of the more troublesome ones.
3. The building up of the herbarium.
4. The preparation of a report upon the flora of the state.

These all received attention to the extent of the time that could be spared from other imperative duties, such as those of the classroom and the routine work of the Station, but it was found necessary to continue the same subjects for 1895. During that time one phase of the first has had attention in Bulletin No. 21, "The Smut of Grains and Potato Scab;" the second in Bulletin No. 19, "Squrrel-Tail Grass (Fox-Tail)," our worst weed, and a Press Bulletin on the "Russian Thistle." The third has, of course, gone on incidentally with and preparatory to the fourth.

Although what has been done in the study of the flora of the state has cost no little time and labor yet the work seems but barely begun. The preparation of a full and reasonably inclusive report on the flora of a great state of nearly 100,000 square miles would be the work of years for a corps of men devoting their full time to the matter in hand, so one man with a full slate of college teaching and other Experiment Station duties, besides that of working up the flora, would lose courage were it not for the absorbing interest of the subject itself. Since to delay the report until it should be approximately complete would project it far into the future and might possibly result in its never being published, it has seemed advisable to publish the results thus far attained. As the work goes on and results accumulate, other reports may, from time to time, appear to record the additions.

COLLECTING TRIPS.

The basis for the following brief report and the catalogue of species rests mainly upon the collections made by the writer in 1894 and in 1895.

With one exception, as given below, no systematic work in collecting had previously been done. In 1892, Prof. B. C. Buffum, at that time acting botanist of the Station, spent the mid-summer months in the field collecting—primarily to secure for the University and the Station a collection of the native grasses and forage plants, an exhibit of which was to be made at the World's Fair at Chicago in 1893. Incidently much more was done, for a considerable amount of good material, other than grasses, must be put down to the credit of the expedition.

From this material were obtained the numbers which formed the nucleus of the present collection in our herbarium. Quite a large part of the state was covered during this trip, extending from Laramie, on the south, to Lander and the Wind River Mountains on the north-west, to Sheridan and the Big Horn Mountains on the north, and Sundance, Fort Laramie, and Wheatland on the east. How many numbers were collected, I am unable to state, as no field or collection numbers were made use of, but in the succeeding catalogue of plants those species where no collection number is noted, are generally to be credited to this expedition. In the Gramineæ and Cyperaceæ much the larger number of species are the result of this earlier expedition in which these groups received such thorough-going attention that in the later collecting trips it seemed advisable to concentrate attention upon the other Phanerogams.

It may be of interest to give briefly the history of the field work that furnished the material upon which this report rests.

1894.

During the continuance of the spring term of school, operations had to be confined to Laramie and adjacent territory. Mornings and evenings, holidays and Saturdays were used with all diligence. The most distant point reached was Table Mountain, about twenty miles to the east, and on the west the Laramie River served as boundary line. Limited as was the area covered and late as seasons are at this altitude, June 30 saw 300 numbers collected and stored in duplicate.

At the annual meeting of the Board of Trustees at the close of the school year, provision was made for an expedition to go into the field during the summer vacation, in the interest of the departments of geology and botany. Preparations for setting out were perfected as rapidly as possible, but it was not until July 7 that the start was made.

Beside the writer

THE PARTY

consisted of W. C. Knight, Professor of Geology in the University and Geologist of the Station; Mr. W. H. Reed, who furnished a large part of the outfit for the expedition; Mr. Geo. M. Cordiner, a student at the University, who accompanied the party as the writer's assistant.*

The expenses of the expedition were reduced to a minimum as they consisted of only the actual living expenses of the party in the field, plus the expense incident to securing the services of Mr. Reed, with his two teams and wagons and one saddle horse. The camp equipage consisted of all the necessary utensils, ample bedding, a tent, which, owing to the perfect weather, was rarely used, a stock of groceries, besides the necessary apparatus for collecting in both botany and geology.

The botanist's outfit may be of interest and was as follows: Two ordinary-sized tin collecting cans, one of which had a number of small compartments at the end for diminutive and delicate objects, and one large tin vasculum so large that it was always referred to as the "tin trunk." This was indispensable during the long mountain trips, when it was a desideratum to be able to bring back once for all a large amount of material. The most efficient instrument for uprooting plants, both on the plains and among the rocks in the mountains, was found

---

It is with deep sorrow that I record the death of this noble young man. On March 13, 1 — he received fatal injuries during a fire in Laramie by being caught under the falling walls of a building, from which he was helping to remove goods. While in the field he greatly endeared himself to the writer by his constant cheerfulness, his remarkable faithfulness to duty, his high efficiency and his moral worth.

to be a carpenter's strong chisel about three-fourths of an inch broad. Of course, a stout knife is not to be omitted from any collecting outfit.

Five plant presses were carried and found none too many at times, but at other times two would have been more than enough. Four of these were slat presses, the pressure in two being applied by straps and in the other two by cords as in the Acme press. The fifth, and most important one, was a screw-press and as it filled its purpose so admirably and proved so valuable an adjunct to the outfit, I add here a figure and brief description.

The base and the top bars are of oak, 2 x 4½ x 20 inches, the distance between them being 21 inches. They are united by one-half inch bolts so placed as to be eighteen inches apart. To keep the base and top in place when the pressure is relaxed strips of board ½x4½ inches are nailed to the ends. The floor to receive the package of plants is 12x18 inches, made by nailing pine board ⅞ inches thick crosswise of the base. The screw is a ¾-inch bolt with 15 threads to the inch, about 15 inches thread bearing. A square nut is sunk on the under side of the top bar and held in position by a piece of tin tacked over it, the tin having a circular hole in it large enough to admit the bolt. The follower is made of a strip of pine 1x5x18 inches, crosswise on which is nailed pine board ⅞ of an inch thick, making the surface 12x18 inches. On the upper face of the follower is fastened a small flat piece of iron with a sunk center to receive the end of the screw. The projecting end of the screw is made square, on which a removable handle with a square slot is used.

This homely, home-made press has the following advantages: It will furnish all the pressure that can possibly be desired; a large quantity of material can be handled at one time; coarse harsh plants can be forced into shape and held there; packages that have been in the screw press for twenty-four hours can be satisfactorily handled in the strap presses; moisture can be forced out so rapidly that by frequent change of driers the drying of the material may be considerably hastened; natural colors can be preserved in plants that by ordinary means would blacken; it furnishes a safe and convenient way for carrying the material while in press and in every way greatly facilitates the making of good flat specimens.

It was found that the capacity of this press is much greater than its dimensions would indicate. With a good quality of carpet felt for driers and single pressing sheets, cut to size (newspaper quality, unprinted), it was possible to do good and rapid work. Experience shows that more plants are spoiled by too little than by too much pressure.

ROUTE AND PRINCIPAL CAMPS.

The accompanying map will indicate the route in a general way but of course cannot show the miles and miles of ground covered in the vicinity of the various camps, nor the wide detours made on horseback during the days the party was traveling. At a rough guess, more than two-thirds of the time was spent on the road, as the distance covered was upwards of 1,000 miles.

The days of travel were busy ones for the Botanist and his assistant, for at noon the collections of the morning were put into press, the driers on all the other recent material were changed, the damp driers thrown out on the hot sand and again collected. When camp was made for the night the afternoon collections were cared for, and, if they chanced to be heavy, the task was often completed by the campfire.

To enter into a detailed account of the journey would take up too much space, possibly to no purpose, so the following skeleton account must suffice: The route from Laramie took a north-east direction over the Laramie Hills past Grant, Wheatland and Uva postoffices to the Platte River, where camp was made July 10. After four days the camp was moved in turn to Hartville, a mining district, to Whalen Canon, and to the Mexican Mines. From this place a side trip was made by the Botanist and his assistant to Lusk. Camp was broken at the Mexican Mines on July 22, and the Platte River again reached at Orin Junction. From this place the river was followed to Bessemer, passing in turn Douglas, Glenrock and Casper. From Bessemer, Poison Spider Creek was followed to its head near Garfield Peak in the Rattlesnake Mountains. Camp for a day was made on Wallace Creek at the foot of the peak. Up to that date, July 29, this was the richest collecting ground so far encountered. The road from there to the Big Popo Agie river lay across a barren desert, whose few streams were dry and the occasional springs mere mud holes, strongly impregnated with alkali.

The evening of July 30, caught us at Alkali Springs, a bog so densely impregnated with salt that only the almost famished horses could swallow a mouthful, and to one or two of the animals the draught came near proving fatal. There was no grass except an occasional stalk of salt-marsh grass—*Distichlis maritima* and *Triglochin maritima*. Thirsty as every member of the party was, after an all day's drive on the burning plain, not even coffee made from such water could be swallowed. Just at sunset a diminutive shower fell, and a dusty wagon sheet was called into service to catch the precious drops. As the little pool formed in the sagging center of the canvas what delicious draughts of nectar, thickened with the dust of many days, were dipped up with the old tin cups.

At four o'clock next morning, breakfastless, the party started for Beaver Creek, twelve miles distant. Breakfasting here where water was good and abundant, even the weakened horses were somewhat revived, though feed was still very scarce. Early in the afternoon the Big Popo Agie was reached by hard effort, and here three days were spent in camp to allow the sick and famished horses to recuperate. Water and grass were both abundant, and the horses soon showed the effect of better treatment.

August 3, the camp was moved to Lander, where a stop of twenty-four hours was made, after which the expedition proceeded to and up the Big Wind River. The route lay across the Shoshone Indian Reservation, one camp being made at Fort Washakie. It was found necessary to ford the Big Wind River thirteen times before Dubois postoffice was reached. This is the last outpost of civilization on the river and it is near this place that the trail over the Wind River Mountains through Union Pass leads off from the river.

The ascent through the pass was begun on August 10, the summit of the range being reached early on the 11th. A most varied and beautiful vegetation was present on every hand all the way up, and at the highest point in the pass, about 9,500 feet, the open parks among the Spruce groves were covered with the wildest profusion of flowers—a veritable botanist's paradise. Two full days were spent here, but that was far too short a time even though the days were considerably lengthened out by the cheerful evening and early morning campfires. Union Peak, with its snow banks, the moist, dense copses on its slopes and the small fertile valleys at its base rivalled each other in the richness of their treasures.

The days were slipping by and the stay could not be prolonged even here, so on the 14th we turned our faces toward the three Teton peaks, whose lofty summits, though yet a hundred miles away, were seen from our present point of vantage standing out against the pellucid sky like mighty spires, their seamed and snow-flecked faces shining with a radiance possible only in such an atmosphere as this. Over trails all but impassable, up hill and down dale, most laboriously we advanced and at last, on the 18th, camp was made in a Cottonwood grove on the banks of the Snake River in Jackson's Hole, near the foot of the Grand Teton.

Here some days were spent, during which an ascent of the Grand Teton was made. On the 20th three of us began the ascent. That night was spent on the shore of a small lake at about 9,500 feet. The next morning the more precipitous slopes were scaled, but at about 11,000 feet further progress was stopped by a frightful chasm which entirely cut off all communication with the spire-like summit that still towered above us. The descent was made more rapidly, and at nightfall on the 21st we were again in camp, loaded down with the collections of that and the preceding day.

This camp marked the most distant point reached by the expedition and on the afternoon of the 22nd the return journey was begun. The route lay up the Gros Ventre River, up Bacon Creek and over the divide separating it from Green River. On reaching Green River we camped for twenty-four hours, after which we proceeded to Cora postoffice on New Fork. At this point the route took a general south-east course, skirting the foot hills on the west of the Wind River range. The following streams were crossed in succession, viz: New Fork, Boulder Creek, East Fork, Silver, Willow, and Muddy Creeks, the Big and the Little Sandy. This brought us to South Pass City, the oldest mining camp in the Wind River Mountains, thirty miles south of Lander.

At this point, on September 1, the writer found it necessary to sever his connection with the expedition and return to Laramie. Mr. Cordiner remained with the party, and, although it was becoming late in the collecting season and the route lay through a rather barren region, made doubly so by the large flocks of sheep that had been driven over the range, yet a number of interesting species were secured, but unfortunately, most of this material was lost by the burning of a car in the Laramie yards the night of its arrival.

The writer was in the field eight weeks and two days and the other members of the party ten weeks and three days. During this extended trip 900 numbers were made (including some that were secured after the return to Laramie), which, added to the 300 collected during the spring, made 1,200 numbers for the season of 1894. These were all collected in duplicate, ten or more sheets of each number being prepared whenever the material could be procured. This omniverous collecting resulted in quite a percentage of duplicates, so that the actual number of species, not counting forms, was probably not much above 1,000.

1895.

In 1895 it was not found possible to spend even the whole of the vacation in the field but all available time was utilized during the entire season. Especial effort was directed toward procuring such species as were not secured in 1894, and it seemed wise to concentrate effort upon a much more limited area. To this end four expeditions were planned and carried out.

The first one left Laramie June 27, going to the east and working Pole Creek, Table Mountain and adjacent territory, 103 species being secured. The second left Laramie July 25, camp being established at Cummins

City, from which point the surrounding mountains and valleys were scoured and resulted in 123 numbers. The third expedition left for Laramie Peak, seventy-five miles distant, on August 3 and returned with 117 numbers. The fourth made two camps, one in the Centennial Valley and one at the La Plata Mines near the summit of the Medicine Bow (Snowy) range. From these two points adjacent territory was worked and yielded 192 species, many of them quite rare.

Besides the above species, there were collected at various times during the season 135 numbers, making a total for 1895 of 670 numbers. As these were all in duplicate, approximately in tens, the total number was near 6,700.

Of the 670 species about one-half are new as compared with the collections of 1894.

PLANT ZONES.

Recent writers have made much of plant zones as limited by given lines of elevation. There is, no doubt, considerable truth in the theory that fairly well marked belts are found, but I think it is possible to overestimate the importance as well as the distinctiveness of such zones. There are so many other factors that enter into the problem, such as moisture, soil and exposure that its solution becomes peculiarly difficult. The zones sink and rise in conformity as much with the configuration of the land, the absence or presence of arboreal vegetation, the character of the soil and the amount of moisture as in respect to the altitude. The monotony of the grassy plain gives place to a veritable garden if but a few clay hummocks or stony points and ravines interrupt its interminable length. Seven thousand feet with one exposure may produce a more truly Alpine flora than 9,000 feet with another.

Many species hold their own at almost all altitudes, and beginning with the lower altitudes, are successively in blossom throughout the season at higher and higher elevations. On the other hand, in given areas, a few certain plants are never met with except within a given range of elevation, but this given elevation differs for different parts of even the same state. It seems to be a a relative point depending as much upon the elevation of the surrounding country as upon the actual elevation above sea-level. It follows, therefore, that plant zones can only be established for a given area, as for instance, the Laramie Plains and the mountains that rise on either hand of it.

It has been well pointed out by Dr. Coville in his report on the "Botany of the Death Valley Expedition" that only certain plants can be taken to mark zonal lines. That only a few comply with the two characteristics of a good zonal plant, viz: "It shall have a definite termination at the borders of a zone or at lines substantially parallel thereto, but closer together." "That the belt of a zonal plant should be continuous." In any region I think this may be found true of a very small number of plants, but the large majority which have to be fitted into these zones will so overlap from zone to zone that no sharp distinctions can be drawn. Of course, between the lowest and the highest zone of a given area the characteristics are quite distinctive. These represent essentially different floras with as little in common as the vegetation of the plains and that of the adjacent mountains.

In this report it has not seemed wise to try to establish the vertical zones on account of the comparatively limited observations yet made within the wide borders of a great state. Rather something may be said of certain areas or characteristic regions.

THE PLAINS FLORA.

The regions referred to as plains differ greatly in respect to soil, rainfall and altitude. All are comparatively level tracts of land devoid of arboreal vegetation, if one excepts the occasional border of Cottonwoods on stream banks. In this report no mention can be made of the plains of the north-eastern, nor of the south-western parts of the state, for these regions are yet to be visited. It is, however, well known that both these regions differ materially from the rest of the state and from each other. The former, with a considerable rainfall and "gumbo" soil; the latter, sandy soil and a minimum of rain. Careful exploration in these two regions will add a large number of species to the list of the state.

The plains from which we have reports, meagre as yet, are the Laramie Plains; the plains lying east of the Laramie Mountains, and south of Lusk; those bordering on the Platte River, and those northward from the Platte through the center of the state to Lander; the plains adjacent to Wind River, and those of the upper part of the Green River valley; also the Gros Ventre valley, and Jackson's Hole. Through this latter mountain-enclosed, plateau-like plain flows Snake River.

All of these may again be classified, either as a whole or in part as: 1. Sandy, or gravelly plains. 2. Alkali plains.

To the first class belong those whose soil is comparatively free from alkali and whose characteristic shrub, when any is present, is the common sage brush (*Artemisia tridentata*). The characteristic undershrub is some form of *Bigelovia*, indiscriminately called White Sage, Rabbit Brush, Golden Rod, etc. Plains of this character may also be denominated grassy plains. The grasses on these of course vary greatly as to the species and the luxuriance of their growth. The following are among those of most frequent occurrence: *Agropyrum glaucum, A. violaceum, Bouteloua oligostachya*, and *B. racemosa, Buchloa dactyloides, Koeleria cristata*, one or two *Festucas* and several *Poas*. Along water courses and in boggy places, as well as in over-irrigated places, these are replaced or become mixed with many species of *Juncus, Scirpus* and *Carex*. Sometimes Foxtail (*Hordeum jubatum*) takes complete possession.

The second class.—The plains strongly impregnated with alkali (sodium carbonate or sodium sulphate), are in some instances nearly devoid of vegetation, but more usually we find several characteristic plants. If the alkali be sodium sulphate the characteristic shrub is *Sarcobatus vermiculatus*, the well-known Grease Wood. On sodium carbonate soil, this, if not replaced by, has mingled with it some form of *Atriplex*, usually *A. confertifolia*, frequently called White Sage. Other species of *Atriplex*, mostly annuals, are found in this character of soil, and if the soil is very strongly impregnated, as on the shores of salt-marshes and partially dried up alkali lakes, the various species of *Atriplex*, of *Sueda* and of *Salicornea* are often the only vegetation. In real alkali bogs we find *Distichlis maritima, Triglochin maritima* and *T. palustre* as the most characteristic vegetation.

The other areas may be spoken of as the foot-hills and the mountains.

FLORA OF THE FOOT-HILLS.

Two kinds of foot-hills must be recognized, viz: wooded and denuded. The denuded slopes are of course much dryer and a large part of the year devoid of all streams. These foot-hills, if stony or gravelly, are covered with *Cercocarpus parvifolius*, *Rhus tridentata*, *Amelanchier alnifolia*, *Purshia tridentata*—one or more in varying proportion. The intervening valleys, if soil is fertile, are usually covered with sage brush. The herbaceous vegetation in such foot-hills is so varied that no list can be offered here, but the following genera are well represented: *Draba*, *Astragalus*, *Potentilla*, *Actinella*, *Erigeron*, *Senecio*, *Krynitzkia*, *Phlox*, *Penstemon*, and *Poa*.

If the soil contains alkali, the above-mentioned shrubs give place to Grease Wood, and the herbaceous vegetation largely disappears.

The wooded foot-hills are less common, but they occur at intervals in the Laramie range, much more frequently in the Medicine Bow Mountains and the Wind River range. The arboreal vegetation consists of only a few species, unless one includes the Willows that skirt most of the streams that flow from the higher mountains. Lodge Pole Pine, Douglas Spruce, Rocky Mountain White Pine, Black Cottonwood (*Populus angustifolia*) and more rarely Blue Spruce, Engelmann's Spruce and Rydberg's Cottonwood (*Populus acuminata*) are the most frequently met with. The shrubs are more varied and include, besides those mentioned for the drier hills, *Juniperus*, *Prunus*, Willows and Quaking Asp. The latter in some places becomes a small tree and is in fact found at all altitudes along streams or on hill-sides below snow banks. The smaller vegetation likewise includes a much greater number of species, each of which apparently strives for the mastery and produces the most beautiful confusion of forms.

THE MOUNTAIN FLORA.

Some of the mountain ranges are quite heavily timbered, notably the Medicine Bow and Wind River ranges. The Laramie Mountains are wooded only in part and some of these areas very sparsely. Other ranges are known to be wooded, but I speak only of those I have visited. The summits of the Laramie Mountains are mostly rounded and undulating, and on these we find a scattering growth of Rocky Mountain Yellow Pine (*Pinus ponderosa scopulorum*) and occasionally some straggling, stunted specimens of the Virginia Juniper. Wherever we find the range broken by more abrupt slopes, deeper canons and water courses, the arboreal vegetation assumes the character of a forest, and in some districts furnishes valuable lumber. This is the case at Laramie Peak and on some of the spurs that run out from it. The forests consist mostly of Douglas Spruce, Rocky Mountain White Pine and Lodge Pole Pine.

Much the larger part of the Medicine Bow Mountains are heavily wooded, and it is from these forests that the larger part of the native lumber used in the southern part of the state is obtained. About the same species prevail as in the Laramie Mountains, with the addition of the Blue Spruce and Engelmann's Spruce. The White Pine (*Pinus flexilis*) and Douglas Spruce form much the larger part of the whole. The latter, along the streams at the foot of the ranges, reaches its greatest size and it gradually comes to form a larger proportion of the whole until at 9,000 feet and upward it constitutes practically an unbroken forest to the exclusion of other species. At timber line it becomes scattering, dwarfed and depressed, spreading out like a huge mat under the enormous pressure of the winter snows.

Practically the same conditions prevail in the Wind River Mountains, and probably, though I cannot speak from observation, in the Big Horn Mountains.

Of the fruticose and herbaceous vegetation I need not speak here, although the summits of these ranges yield many beautiful and strictly alpine forms. These all receive comment in their proper place in the list, so space may not be consumed for that purpose here.

THE TREES OF THE STATE.

A list of the trees of the state is indeed very short and were those on the border line between trees and shrubs excluded in would be shorter yet by a third.

Rocky Mountain Yellow Pine (*Pinus ponderosa scopulorum*).
Rocky Mountain White Pine (*Pinus flexilis*).
Lodge Pole Pine (*Pinus Murrayana*).
Engelmann's Spruce (*Picea Engelmanni*).
Blue Spruce (*Picea pungens*).
Douglas Spruce (*Pseudotsuga Douglasii*).
Virginia Juniper (*Juniperus Virginiana*).
Black Cottonwood (*Populus angustifolia*).
Rydberg's Cottonwood (*Populus acuminata*).
Quaking Asp, Aspen (*Populus tremuloides*).
Willow (*Salix longifolia*).
Willow (*Salix flavescens*).
Willow (*Salix amygdaloides*).
Willow (*Salix lasiandra*).
Green Ash (*Fraxinus viridis*).
· Box Elder (*Negundo aceroides*).

Scrub Oak (*Quercus undulata*).
Wild Plum (*Prunus Americana*).
Wild Cherry (*Prunus demissa*).
Wild Cherry (*Prunus Virginiana*).
Hawthorn (*Cratægus rivularis*).
Hawthorn (*Cratægus Douglasii*).
Service Berry (*Amelanchier alnifolia*).
(*Eleagnus argentea*).
Buffalo Berry (*Shepherdia argentea*).
Black Birch (*Betula occidentalis*).
Black Alder (*Alnus incana virescens*).
Sage Brush (*Artemisia tridentata*).

In a few localities of the state occasional specimens of Sage Brush attain a remarkable size—small trees in fact—so that a man on horseback may ride erect underneath the branches.

Other species have been reported but until the specimens are at hand they will not be listed. The number of shrubby plants is so great that to list them separately would be to reprint a large part of the succeeding systematic list.

THE FLORAS OF THE PACIFIC AND THE ATLANTIC SLOPES.

I have had no opportunity to compare the floras of the two regions except in so far as the plants on the western side of the Wind River range and those of Jackson's Hole and the Tetons, all of which are on the Pacific slope, may be compared with those on the eastern side of the Wind River Mountains and those of the south-eastern part of the state, which represent the Atlantic slope. Such examination has led to the conclusion that the continental divide, though dividing the waters, does not separate floras. The two regions have a far larger number in common than they have of forms that are distinct. While my collection lists show much that is different, I firmly believe that the difference is due mainly to the season in which each was collected. I am confirmed in this by an examination of the lists given in Dr. J. M. Coulter's report on the Botany of the Hayden U. S. Geological Survey, 1872. These lists comprise those collected, 1. On both slopes; 2. Only on the eastern slope, and 3. Only on the western slope. Those of the last list were collected in the earlier part of the season, and a remarkable number are the same as those of my list for the eastern slope during the same months. Different localities and seasons yield different results, but complete collections would reveal no abrupt transitions; Iowa and Utah, for instance, have different floras, but any fifty or one hundred miles between, even at the summit of the Rockies, will show only the most gradual substitutions. One form disappears, a new one appears, but this occurs with a change of locality in any direction. On the plains of the Platte and its tributaries *Cleome integrifolia* only is found, while on Wind River and its tributaries *Cleome lutea* is the only form.

INTRODUCED PLANTS.

By introduced plants reference is made only to such as grow without cultivation. Most of them may properly be called "Weeds." This is an ever-growing list and will soon include a large part of those familiar to the eastern farmer. The extension of our agricultural interests of course includes the importation of seed, and rarely is any kind of seed free from weed seeds. Some of our weeds, however, are native plants and thrive immensely under cultivation. Those of special interest receive notice in the proper place in the list, and possibly a future bulletin may deal with the weed problem in this state.

HARDINESS OF NATIVE PLANTS.

The power to withstand frost, so remarkably developed in mountain floras, has undoubtedly often been remarked upon before, but it is, nevertheless, unceasingly a cause of wonder. To see great beds of Phlox, Mertensia, Gilia, Actinella and scores of others in full bloom at times when the temperature at night is $5°$ to $20°$ F., below the freezing point is a phenomenon that can scarcely be explained. That reduced atmospheric pressure plays an important part in preventing injury, I think must be accepted, for the same plants at lower altitudes would perish. The following observation goes to prove this: Late in August in 1890, a plot of potatoes was noted in full blossom at Mountain Home, elevation about 9,000 feet. Observations on three successive days showed no trace of injury though on both of the intervening nights there were heavy white frosts and films of ice formed on water pails. Such a degree of cold would have absolutely killed potatoes at sea-level.

FLORAL CALENDAR.

In 1894 every effort was made to keep pace with the floral procession. At the altitude of the Laramie Plains, (7,000 feet), Spring opens comparatively late. April furnishes very few objects of interest to the botanist. The earliest flowers are *Phlox cæspitosa* and *Townsendia sericea*, both of which expand their blossoms scarcely above the surface of the ground. These are soon followed by some small *Umbelliferæ*, among which *Cymopterus montanus* may be noted. Toward the end of the month a few more begin to appear on the plains and in the foot-hills, all of which possess either large, fleshy, perennial roots, as *Leucocrinum montanum*, *Musenium trachycarpum* and *Peucedanum nundicaule*, or else they have large woody subterranean stems from which spring the small leaves and numerous flowers that spread out in dense mats or cushions upon the cold soil. Such are *Astragalus spatulatus* and *Astragalus sericoleucus*. Among the rocks in sheltered nooks are also two *Mertensias*, *lanceolata* and *alpina*. With the advent of May, or sometimes earlier, the little *Drabas*, *glacialis* and *alpina* tinge the naked rocks with yellow. In the moister canons our earliest Buttercup, *Ranunculus glaberrimus*, and the Wind-flower, *Anemone patens Nuttalliana*, are found.

Very slowly through May, for cold days and snow-storms are far from rare, the number grows so that the diligent observer may find several score. From this time on the forms crowd upon each other in rapid succession and one soon loses track of the order of their coming. June is the floral month of the plains, July of the lower mountains and August is the month of months in the high altitudes. September has something of worth everywhere and a few forms linger late into October.

BOTANICAL WORK IN THE STATE.

So far as I have been able to learn there are no other workers in systematic botany in the state, nor are there any other herbaria, public or private. On this account I have been unable to make comparison of our specimens with those from other localities in the state.

As before stated the following list is based wholly upon specimens in our herbarium. Plants reported as in the state are not included but appended in separate lists, each under heads showing by whom and from what locality reported.

NOMENCLATURE AND CITATIONS.

As the discussion of the nomenclature question is still waxing warm, happily with less acerbity than before, it has seemed almost a matter of indifference what view of the question was taken with reference to this report. Not that it is a matter of indifference to any worker in the field of botany, but the question seems so far from settlement that one may still expect almost any solution. Stability is the object all have in view and those who publish work of any kind will use that nomenclature which seems to them to offer the greatest chances for permanency. So long as different adherents of the so-called "new nomenclature" are far from agreed among themselves, as witness recent publications, there is small inducement to abandon a fairly satisfactory system. That there is room for improvement none will deny, but until there is international agreement we shall hardly reach a permanent, much less an ideal nomenclature.

I greatly regret that citations in many instances are quite incomplete. Meager library facilities must be my excuse.

Where the recent "List of Pteridophyta and Spermaphyta of the Northeastern United States" recommends a name, different from that used in this report, such name is given as a synonym without citation.

For the citations given I am often indebted to the

above list; to recent publications from the Division of Botany, U. S. Dept. of Agriculture; to various publications from the Gray Herbarium and to several other papers and reports.

CRYPTOGAMS.

In 1894 no Cryptogams were collected except a few Ferns. In 1895 efforts were made to secure the Mosses as well. No attention was given to the other groups but incidentally a few Lichens and Fungi were picked up. In future collecting it is purposed to give more attention to this part of the flora.

ACKNOWLEDGEMENTS.

I have pleasure in acknowledging the generous assistance of the various members of the collecting party of 1894. To Prof. W. C. Knight I am indebted for specimens and information in the field and subsequently for literature and data upon various topics. The services rendered by Prof. B. C. Buffum in the field I have previously mentioned; his continued interest and occasional assistance is greatly appreciated. To Mrs. Celia A. Nelson much credit is due for her painstaking care in the preparation of specimens and her devotion to the work during the expeditions of 1895. The task of putting this manuscript in shape for the printer is work for which I am also largely indebted to her.

For assistance in the determination of certain groups I am greatly indebted to the following specialists: To Dr. B. L. Robinson, to whom a considerable number were submitted for comparison and determination. For careful reports upon these I am indebted to him and Mr. M. L. Fernald. Certain orders and genera were submitted only in part as follows: *Astragalus* to Prof. E. P. Sheldon; *Umbelliferæ* to Dr. J. N. Rose; *Juncus* to Dr. F. V. Coville; *Gramineæ* to Dr. F. Lamson-Scribner; *Carex* to Prof. L. H. Bailey; *Salix* to the late Mr. M. S. Bebb; *Filices* to Prof. L. M. Underwood; *Musci* to Prof. J. M. Holzinger; *Fungi* to Mr. J. B. Ellis; *Lichens* to Dr. J. W. Eckfelt. Lastly I would mention the kindness of Prof. E. L. Greene, who has furnished determinations, corrections and valuable suggestions on a number of miscellaneous specimens.

EXPLANATIONS.

All comments upon plants, such as rare, frequent, etc., refer to this state and in partictular to those parts of the state which this report covers. The number in parrenthesis following the comments upon each species is the collection number and will render reference to the specimen easy in case any are found incorrectly determined.

PRINCIPAL CAMPS AND COLLECTING LOCALITIES.

The following table contains the principal points in the neighborhood of which most of the material was secured. The altitudes given are only approximate and for hills, mountains, etc., the extremes may be less or greater than given.

NO.	PLACE.	COUNTY.	ELEVATION—FT.
1.	Alkali Springs	Fremont	5500
2.	Bacon Creek	Uinta	7000-8000
3.	Bessemer	Natrona	5300
4.	Big Muddy Creek	Converse	3000
5.	Big Popo Agie River	Fremont	5200
6.	Big Wind River	Fremont	5200
7.	Blue Grass Creek	Albany	6000-7000
8.	Blue Grass Hills	Albany	6500
9.	Boulder Creek	Fremont	7400
10.	Bull Lake	Fremont	5350
11.	Bull Lake Creek	Fremont	5300
12.	Casper	Natrona	5200
13.	Casper Mountain	Natrona	5500-6500
14.	Centennial Hills	Albany	8000-9500
15.	Centennial Valley	Albany	7500-8000
16.	Clark's Ranch	Fremont	7200
17.	Cottonwood Canon	Albany	6000-6500
18.	Cummins, vicinity of	Albany	8000-9000
19.	C. Y. Ranch	Converse	5000
20.	Dubois	Fremont	7000
21.	East Fork	Fremont	7300
22.	Fairbanks	Laramie	4200
23.	Ford J. Ranch	Laramie	4500
24.	Fort Washakie	Fremont	5200
25.	Garfield Peak	Natrona	6000-8000
26.	Grant	Laramie	6000
27.	Green River	Uinta	8000
28.	Gros Ventre River	Uinta	7000
29.	Hartville	Laramie	4500
30.	Horse Creek	Albany	7300-7800
31.	Howell Lakes	Albany	7000
32.	Jackson's Hole	Uinta	6500
33.	Jelm Mountain	Albany	8000-9000
34.	Lander	Fremont	5300
35.	La Plata Mines	Albany	11000
36.	Laramie	Albany	7300
37.	Laramie Hills	Albany	7500-8500
38.	Laramie Peak	Albany	7000-9000
39.	Laramie Plains	Albany	7000
40.	Little Sandy	Fremont	7000
41.	Lusk	Converse	5000
42.	Meadow Creek	Fremont	6500
43.	Mexican Mines	Converse	4750
44.	Muddy Creek	Fremont	7000
45.	Musk-rat Creek	Fremont	5300
46.	Platte River	Laramie	4200
47.	Poison Spider Creek	Natrona	5500-6000
48.	Pole Creek	Albany	7000-8000
49.	Seven-mile Lake	Albany	7000
50.	Silver Creek	Fremont	7000
51.	South Pass City	Fremont	7500
52.	Sweetwater River	Fremont	6000-7000
53.	Sybille Creek	Albany	6000-7000
54.	Table Mountain	Laramie	7000-8000
55.	Telephone Canon	Albany	7500-8000
56.	Teton Mountains	Uinta	7000-11000
57.	Union Pass	Fremont	8000-10000
58.	Union Peak	Fremont	10000-11000
59.	Uva	Laramie	4500
60.	Warm Spring Creek	Fremont	8000
61.	Whalen Canon	Laramie	4750
62.	Wheatland	Laramie	4700
63.	Willow Creek	Laramie	4500
64.	Wind River, North Fork of	Fremont	6500

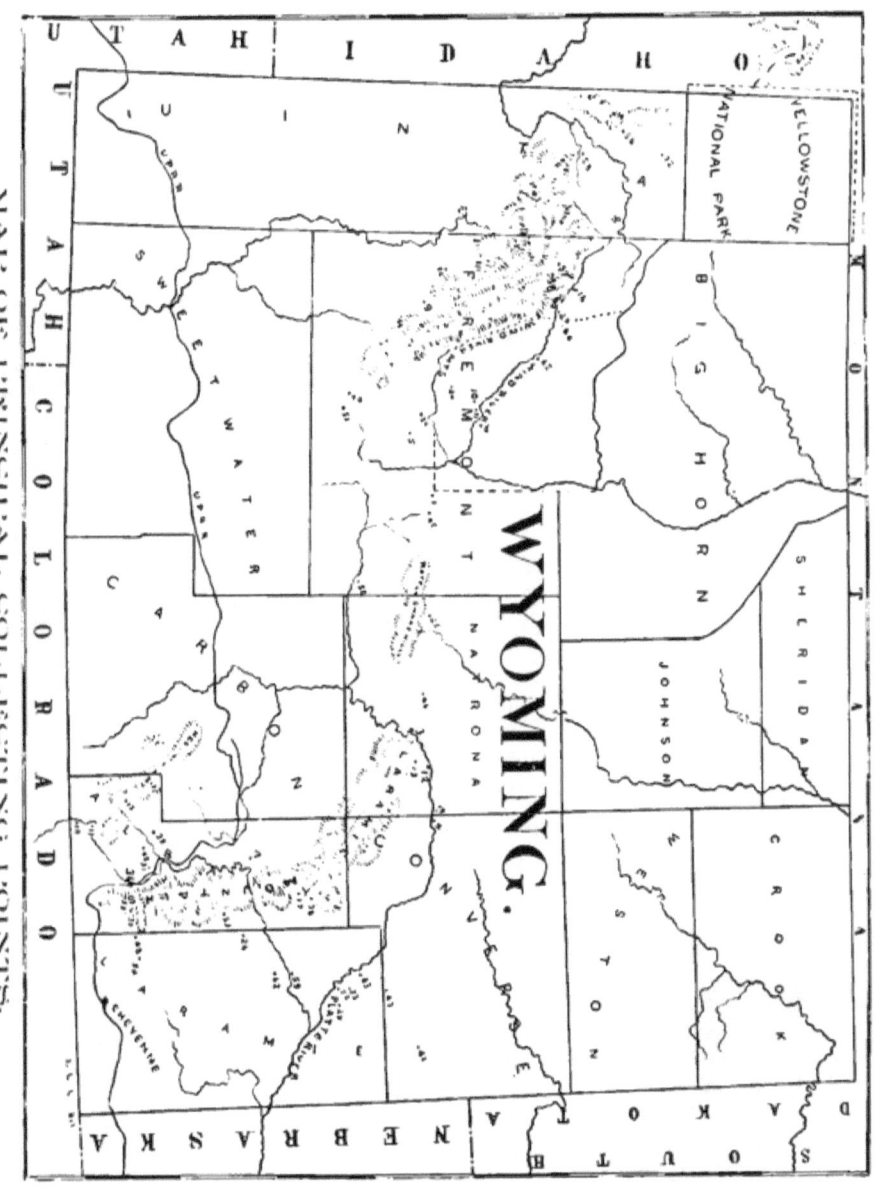

LIST OF SPECIES.

RANUNCULACEÆ.

Clematis ligusticifolia, Nutt, Torr. & Gray, Fl. i, 9 (1838).
From type locality, viz: "Plains of the Rocky Mountains, in open and in bushy places near streams." Rather frequent.
Blue Grass Creek, Fl. July 8, 1894 (No. 361); Popo Agie River, Fr. August 3, 1894 (No. 712). *Virgin's Bower*.

Clematis Douglasii, Hook. Fl. Bor. Am. i, 1 (1829).
Observed only in one locality, on open hillside.
Laramie Hills, June 9, 1894 (No. 202).

Clematis verticillaris Columbiana, M. E. Jones.
Infrequent, on moist wooded hillsides.
Pole Creek, May 25, 1894 (No. 72).

Thalictrum Fendleri, Engelm. Gray Pl. Fendl. 5.
Common in wet Meadows.
Horse Creek, June 9, 1894 (No. 189); East Fork, August 25, 1894 (No. 1119). *Meadow Rue*.

Thalictrum occidentale, Gray. Proc. Am. Acad. viii, 372.
In thickets along mountain streams, variable as to number of akenes which sometimes are 7-12.
Jackson's Hole, August 20, 1894 (No. 937); Gros Ventre, August 25, 1894 (No. 1067).

Thalictrum purpurascens, L. Sp. Pl. 1753.
Found but once, probably confined to the north and east.
Whalen canon, July 17, 1894 (No. 512);

Thalictrum sparsiflorum, Turcz. in Ind. Sem. Petr. i, 40.
On mountain streams, 8000 ft. and upward.
Cummins, July 30, 1895 (No. 1494).

Anemone cylindrica, Gray. Ann. N. Y. Lyc. iii, 221.
Not common, in fertile valleys, near streams.
Laramie Peak, August 5, 1895 (No. 1583).

Anemone multifida, Poir, Suppl. i, 364.
    Infrequent, moist mountain valleys.
        Laramie Hills, June 22, 1894 (250), B. C. Buffum ; Upper Wind River, August 10, 1894 (No. 756).
Anemone patens Nuttalliana, Gray, Man. Ed. 5, 36 (1867). *Pulsatilla hirsutissima,* (Pursh) Britton.
    One of the earliest flowers, on moist. rich canon-sides.
        Laramie Hills, Fl. May 4, 1894 (No. 2); Fr. May 29. *Wind Flower.*
Anemone Pennsylvanica, L. Mant. ii. 247. *A. Canadensis,* L.
    Very abundant on streams in the eastern part of the state.
        Pole Creek, June 2, 1894 (No. 139); Sybille Creek, July 8, 1894 (No. 406).
Ranunculus acriformis, Gray, Proc. Am. Acad. xxi, 374.
    Common on creek and river bottoms, at least in Albany county.
        Horse Creek, June 9, 1894 (No. 199); Little Laramie River, June 9, 1895 (No. 1306). *Buttercup.*
Ranunculus affinis cardiophyllus, Gray, Proc. Acad. Phila. 56 (1863).
    This seems to be a good variety, easily distinguished from var. *validus,* Gray, by the smaller size of the plant, and by the thinner leaves which are either somewhat acute or show a tendency to become divided. Mountain meadows at 7000-8000 ft. Pole Creek, June 2, 1894 (No. 126).
Ranunculus affinis validus, Gray.
    Wet meadows, Horse Creek, June 9, 1894 (No. 195).
Ranunculus alismæfolius, Geyer. Benth. Pl. Hartw. 295 (1839).
    This is not quite typical and is probably var. *montanus,* Watson, or *R. alismellus,* Greene. Medicine Bow Mountains, at 11,000 ft. August 23, 1895 (No. 1762).
Ranunculus aquatilis trichophyllus, Gray, Man. Ed. 5, 40 (1867). *Batrachium trychophyllum,* (Chaix) Bossch.
    Common in slow-moving streams. Muddy Creek, August 25, 1894 (No. 1112).
Ranunculus Cymbalaria, Pursh, Fl. Am. Sept. 392 (1814). *Cyrtorhyncha Cymbalaria,* (Pursh) Britton.
    Very common in wet alkali soils. Laramie, at various times, the plant remaining in blossom throughout the season; Lander, August 3, 1894 (No. 693).

Ranunculus Eschscholtzii, Schlecht. Animad. Ranunc. ii, 16.
Union Peak, August 13, 1894 (No. 1003); Medicine Bow Mountains, at 11,000 ft., August 22, 1895 (No. 1780).

Ranunculus eximius, Greene, Erythea iii, 19 (1895).
A most beautiful large-flowered species collected by B.C. Buffum at Bald Mountain, August 15, 1892. The specimens in our herbarium are in part the ones from which the original description was drawn.

Ranunculus flammula reptans, Mey. Pl. Lab. 96 (1836). *R. reptans,* L.
Not common, creeping among the stones on the shore of Bull Lake Creek, August 9, 1894 (No. 729).

Ranunculus glaberrimus, Hook. Fl. Bor. Am. i, 12, t. 5.
Our specimens are the form represented by Prof. Greene's *R. ellipticus,* Pittonia, ii, 110. Our earliest Buttercup; very abundant among the sage brush in moist valleys; Laramie Hills, April and May, 1894 (No. 3).

Ranunculus Macounii, Britton, Trans. N. Y. Acad. Sci. xii, 3 (1892).
Frequent on low-lying wet lands; Big Wind River, August 9, 1894 (No. 723); Cummins, July 29, 1895 (No. 1483).

Ranunculus natans, C. A. Mey. in Ledeb. Fl. Alt. ii, 315.
In the mountains, on the muddy banks or in the water of partially dried up lakes. Union Pass, August 10, 1894 (No. 808); East Fork, August 27, 1894 (No. 1113).

Ranunculus Nuttallii, Gray, Proc. Acad. Phila. 56 (1863). *Cyrtorhyncha ranunculina,* Nutt.
Very abundant on rocky ridges at 8,000-9,000 ft. Laramie Hills, May and June, 1894 and 1895 (Nos. 76 and 1237).

Ranunculus Purshii, Richards. Frank. Journ. 741 (1823).
In the bed of a recently dried up lake, Union Pass, August 14, 1894 (No. 880).

Ranunculus rhomboideus, Goldie. Edinb. Phil. Journ. vi, 329 (1822). *R. ovalis,* Raf.
Rare, observed only in wet meadows on Pole Creek, May 25, 1894 (No. 78).

Ranunculus sceleratus, L. Sp. Pl. 551 (1753).
Abundant in shallow spring pools; Fairbanks, July 14, 1894 (No. 453).

Caltha leptosepala, D. C. Syst. i, 310.

Very abundant in wet grounds at 9,000-10,000 ft., greedily eaten by elk and locally called "Elk Slip." Union pass, August 13, 1894 (No. 1023).

Trollius laxus, Salisb. Trans. Linn. Soc. viii, 303 (1803).

In the mountains at 9,000 ft. and upward. Union Pass, August 14, 1894 (No. 886); Medicine Bow Mountains, August 22, 1895 (No. 1709).

Aquilegia cærulea, James. Long's Exped. ii, 345.

The queen of Columbines, superbly handsome. In the woods at 8,000-10,000 ft. Laramie Hills, June 22, 1894 (No. 249), B. C. Buffum.

Aquilegia cærulea alpina, n. var.

This, it seemed to me, must be *A. pubescens*, Coville, but Dr. Robinson thinks it is rather a form of *cærulea*. Further examination convinces me that this is right. It is hardly Dr. Gray's var. *albiflora*, for these specimens are all distinctly yellow and in habitat strictly alpine. The variety differs from the species in the smaller size of the plant and larger leaves with upper leaflets entire; in the smaller flowers and very much shorter spurs.

Possibly confined to the Wind River Mountains, where it occupies crevices and ledges of the naked summits above timber line. Observed by Prof. B. C. Buffum, in such locations in 1892 and collected by the writer on Union Peak at 10,500 ft., August 13, 1894 (No. 894).

Aquilegia Laramiensis, n. sp.

Many stemmed from a rather large, semi-fleshy root; 6-9 inches high. The stems and petioles inclined to be decumbent and diffuse. Spurs short, hooked and knobbed. Slightly pubescent on the underside of leaves, on flowers, follicles and pedicels. Sepals greenish white, lanceolate with emarginate apex; lamina, of the light cream-colored petals, obtuse elliptical, longer than the spur.

It differs from *A. saximontana* in its greater pubescence and larger leaflets; from *A. brevistyla* in having longer petioles with dilated bases; from *A. flavescens* in habit and especially in habitat.*

---

*I am indebted to Dr. Robinson for making comparison of this with the nearly related species.

Collected at the foot of Laramie Peak, in the Laramie range, in a canon where it occupied the dry crevices in abrupt cliffs.

Delphinium azureum (?) Michx. Fl. Bor. Am. i, 314 (1803). *D. Carolinianum*, Walt.

Not quite typical but probably a form of this species. Saratoga, June 1893. J. D. Parker.

Delphinium bicolor, Nutt. Journ. Acad. Phila. vii, 10.

Our earliest Larkspur, in dry loam soil of ravines and hillsides. Telephone Canon, May 23, 1894 (No. 48); Table Mountain, June 28, 1895 (No. 1385).

Delphinium Geyeri, Greene.

A good species, immensely abundant on the Laramie Plains. Frequently greedily eaten by hungry cattle with fatal results, caused by bloating, hence the local common name, "Poison Weed." Laramie Hills, July 10, 1894 (No. 400); South Laramie Plains, July 31, 1895 (No. 1552).

Delphinium scopulorum glaucum, Gray. Bot. Gaz. xii, 52. *D. glaucum*, Wats.

In the mountains at 8,000-9,000 ft.; Union Pass, August 10, 1894 (No. 874); Laramie Peak, August 7, 1895 (No. 1599).

Delphinium scopulorum subalpinum, Gray. Bot. Gaz. xii, 52.

A very beautiful alpine form. LaPlata Mines, 11,000 ft., August 22, 1895 (No. 1761).

Aconitum Columbianum, Nutt. T. & G. Fl. i, 34.

Two marked forms occur: The typical one, tall, large-leaved and dark blue or purple flowered; the other, repeatedly observed, has light yellow flowers, the plant is smaller, leaves smaller and less pubescent. Less striking differences have sometimes been held to be specific, so it seems proper that this at least bear the varietal name *ochroleucum*. The former is common in thickets on mountain streams. Collected at Snake River, August 21, 1894 (No. 939); Cummins, July 29, 1895 (No. 1521). The latter, less frequent, in similar locations; collected at Cummins and in Centennial Valley and observed at Laramie Peak.

Actea spicata arguta, Torr. Pacif. R. Rep. iv, 63.

White and red berried forms occasionally found growing together. Garfield Peak, July 29, 1894 (Nos. 681 and 692); Cummins July 27, 1895 (No. 1490).

BERBERIDACEÆ.

Berberis Repens, Lindl. Bot. Reg. t. 1176 (1828).
    Common among the rocks in hilly regions.   Laramie Hills, May 25, 1894 (No. 66); Centennial Hills, August 1895.

NYMPHÆACEÆ.

Nuphar polysepalum, Engelm. Trans. St. Louis Acad. ii, 283 (1865).
    Nymphœa polysepala, (Engelm.) Greene.
        Sub-alpine ponds and lakes, not common.
        Medicine Bow Mountains, by Col. S. W. Downey; Union Pass, August 14, 1894 (No. 898).   *Yellow Pond Lily.*

PAPAVERACEÆ.

Argemone platyceras. Link & Otto, Icon. i, 85. t, 43 (1828). *Argemone albiflora*, Hornem.
    Common on the eastern slopes of the Laramie Hills.
    Chugwater Station, B. C. Buffum, July 14, 1891; Sybille Hills, July 8, 1894 (No. 314); Table Mountain, July 1, 1895 (No. 1363). *Poppy.*

FUMARIACEÆ.

Corydalis aurea, Willd. Enum 740 (1809). *Capnoides aureum*, (Willd) Kuntze.
    This and the following occurs frequently in dry sandy ravines and on the adjoining hillsides.
    Centennial Valley, August 19, 1895 (No. 1679).
Corydalis aurea occidentalis, Engelm.
    The variety is far more common than the species.
    Telephone Canon, June 3, 1894 (No. 118); Laramie, June 5, 1895 (No. 1241).

CRUCIFERÆ.

Nasturtium obtusum, Nutt. T. & G. Fl. N. A. i, 74 (1838). *Roripa obtusa* (Nutt.) Britton.
    Usually found growing in spray of minature waterfalls.
    East Fork, August 25, 1894 (No. 1116); Pole Creek, July 2, 1895 (No. 1415).

Nasturtium officinale, R. Br. in Ait. Hort. Kew. Ed. 2, iv, 110 (1812). *Roripa Nasturtium.* (L.) Rusby.

Not common, but introduced into some of the springs about Laramie and probably elsewhere in the state.

State Fish Hatchery, September 24, 1894 (No. 1152). *Water Cress.*

Nasturtium palustre, DC Syst. Veg. ii, 191 (1821). *Roripa palustris.* (L.) Bess.

Only rarely seen, wet lowlands.

C. Y. Ranch, on Big Muddy, July 24, 1894 (No. 638).

Nasturtium palustre hispidum, Fisch & Mey. Ind. Sem. Petr. iii, 41. *Roripa hispida* (Desv.) Britton.

This was found twice but only a specimen or two at a time.

Sybille Creek, July 9, 1864 (No. 300.); South Pass, September 2, 1894 (No. 1185).

Nasturtium sinuatum, Nutt. T. & G. Fl. N. A. i, 73 (1838). *Roripa sinuata,* (Nutt.) A. S. Hitchcock.

Apparently quite common, a weed on the campus and on the Experiment farm. (No. 281).

Nasturtium sp.?

A single specimen collected by B. C. Buffum at Bald Mountain, no fruit. Its affinities are not readily made out at this stage but evidently is none of the above.

Barbarea vulgaris, R. Br. in Ait. Hort. Kew, Ed. 2, iv, 109 (1812). *B. Barbarea,* (L.) Mac M.

Union pass, August 10, 1894 (No. 864); Stout plants, thick pods but otherwise normal.

Arabis.

(For this and a few other genera see appendix.)

Thelepodium. (See appendix.)

Cardamine Breweri, Watson, Proc. Am. Acad. x, 339.

Common, at the water's edge in many of our streams. Pole Creek, June 2, 1894 (No. 158); Cummins, July 28, 1895 (No. 1465).

Cardamine cordifolia. Gray, Pl. Fendl. 8.

On stream banks in mountain thickets, quite infrequent. Cummins, July 27, 1895 (No. 1488).

Cardamine Pennsylvanica, Muhl. Sp. Pl. 3: 486 (1800).

Found but once, Lander Creek, August 30, 1894 (No. 1106).

Lesquerella Ludoviciana, Watson, Proc. Am. Acad. xxiii, 252 (1888). *Vesicaria Ludoviciana,* D. C.

Remarkably abundant on the Laramie Plains, in dry sandy soil. June 1, 1894 (No. 190).

Lesquerella montana, Watson, l. c. 251. *Vesicaria montana,* Gray.

Very abundant on the sandy, stony foothills of the Laramie range. Table Mountain June 2, 1894 (No. 88); Pole Creek, June 30, 1895 (No. 1370).

Physaria didymocarpa, Gray, Gen. Ill. i. 162 (1848).

In gravelly clay banks, infrequent, not readily distinguished from Lesquerella until the pods begin to mature. Gros Ventre River, August 16, 1894 (No. 927); Cummins, July 30, 1895 (No. 1555).

Draba. (See appendix.)

Draba alpina, L. Sp. Pl. ii. 642 (1753).

Abundant, on stony gravelly ridges on the plains and in the foothills. Laramie Hills, May 4, 1894 (No. 4); Laramie River Divide, June 9, 1895 (No. 1223).

Draba crassifolia, Graham, Edinb. New Phil. Journ. 182 (1829).

Infrequent, damp, shaded ground; LaPlata Mines, August 21, 1895 (No. 1838).

Draba glacialis, Adams, Mem. Soc. Nat. Mosc. v, 106.

Frequent and of similar habitat as *D. alpina.* Maturing fruit early in May. (Nos. 62 and 1218).

Draba stenoloba, Ledeb. Fl. Ross. i, 154 (1841).

Very rare, collected by B. C. Buffum, in a gulch near Bald Mountain, August 15, 1892.

Sisymbrium canescens, Nutt. Gen. ii, 68.

As variable as it is frequent in occurrence. Table Mountain, June 30, 1895 (No. 1321); Laramie, June 16, 1894 (No. 247).

Sisymbrium incisum, Engelm. Gray Pl. Fendl. 8.

Difficult to separate from some of its varieties. Wheatland, July 9, 1894 (No. 473); South Pass, August 31, 1894 (No. 1184).

Sisymbrium incisum filipes, Gray, Pl. Fendl. 8.

Not at all common; Laramie, August 1893; near Table Mountain, June 30, 1895 (Nos. 1349 and 1425).

Sisymbrium incisum Hartwegianum, Wats. Bot. cal. i, 41.

Observed but once; Sand Creek, August 26, 1894 (No. 1100).

Sisymbrium linifolium, Nutt. T. & G. Fl. i, 91.

Abundant and of frequent occurrence on the plains and in the foothills. Laramie, June 12, 1894 (No. 173); June 1895 (No. 1420).

Smelowskia calycina, C. A. Meyer, Ledeb. Fl. Alt. iii, 165.

Very rare, found but once and then but a plant or two; Laramie Hills, June 1893.

Erisymum asperum Arkansanum, Gray. Man. Ed. 5, 69.

This variety is of very frequent occurrence; in early summer it forms a very conspicuous object on the sandy plains and hillsides in Albany county. Table Mountain, June 2, 1894 (No. 87); Two Bar Ranch on Blue Grass Creek, July 9, 1894 (No. 377).

Erisymum cheiranthoides, L. Sp. Pl. 661 (1753).

Widely distributed but not particularly abundant; Sybille Creek, July 8, 1894 (No. 404); South Pass, September 1, 1894 (No. 1187), and Cummins, July 30, 1895 (No. 1462).

Erisymum parviflorum, Nutt. T. & G. Fl. N. A. i, 95 (838.) *E. inconspicuum.* (Wats.) MacM.

Frequent; Laramie, June 20, 1894 (No. 221); Bacon Creek, August 15, 1894, (No. 916).

Stanleya pinnatifida, Nutt. Gen. ii, 71 (1818). *S. pinnata,* (Pursh) Britton.

Only occasionally, on dry gravelly banks; Little Laramie River, June 6, 1894, by Mr. Houghton; Wood's Landing, July 31, 1895 (No. 1554).

Stanleya pinnatifida integrifolia, Robinson, Syn. Fl. i, 179 (1895).

This form much less common; stems numerous from a large deep-set root; on the dry plains west of Laramie, August 26, 1895 (No. 1845).

Stanleya viridiflora, Nutt. T. & G. Fl. i, 98.

Very rare, not found at all by the writer; collected at Wheatland, June 16, 1892, B. C. Buffum.

Brassica campestris, L. Sp. Pl. 666 (1753).

Occasionally seen in waste places about town; Laramie, September 15, 1893.

Brassica sinapistrum, Boiss. Voy. Espagne, ii, 39 (1839-45).

A single specimen from Centennial Valley, August 25, 1895 (No. 1876).

Capsella Bursa-Pastoris, Medic. Pfl. Gatt. i, 85 (1792). *Bursa Bursa-Pastoris*, (L.) Weber.

A weed in lawns and dooryards everywhere; University Campus, June 1891.

Lepidium apetalum, Willd. Spec. iii, 439.

Quite common on the Laramie Plains, in some places becoming a weed.

Blue Grass Hills, July 8, 1894 (No. 323); University Campus, July 22, 1895 (No. 1424).

Lepidium montanum, Nutt. T. & G. Fl. i, 116, 669.

Rare, State Fish Hatchery grounds, Laramie, B. C. Buffum, 1892; Carbon, June 18, 1894 (No. 257), Miss Lily Boyd.

Lepidium Virginicum, L. Sp. Pl. 645 (1753).

Occasionally found introduced into lawns and vacant lots. Laramie, June 15, 1891, B. C. Buffum.

Thlaspi alpestre, L. Sp. Pl. ii, 903.

The typical form of this species is abundant in the Laramie Hills on open hillsides at 7,000-8,500 ft. Telephone Canon, May 21, 1891, B. C. Buffum; Pole Creek, May 12, 1894 (No. 28).

Thlaspi alpestre glaucum, n. var.

The perennial basal part of stem rather freely branched, herbaceous stems simple and erect, 6-10 inches high; radical leaves broadly to narrowly elliptical, entire or obscurely repand-denticulate; cauline deltoid-auriculate entire. It also differs from the species in the glaucus hue of the leaves, the laxer inflorescence and well marked notch at the apex of the capsule as well as in its habitat. The species flowers in early spring on open hillsides; the variety was collected in the forest almost at timber line, growing in the thick beds of Spruce needles. La Plata Mines, August 21, 1895 (No. 1777).

CAPPARIDACEÆ.

Cleome integrifolia, T. & G. Fl. i, 142.

An obnoxious weed, sometimes occupying acres of ground to the exclusion of everything else. Everywhere on the Laramie plains, and, in fact, all over the south-eastern part of the state. Laramie, June 24, 1894 (No. 297).

Cleome lutea, Hook. Fl. Bor. Am. i, 70, t. 25.

This seems to replace the preceding on the Wind Rivers and in the north-west generally; in sandy soil. Big Wind River, August 5, 1894 (No. 701).

Polanisia trachysperma, T. & G. Fl. N. A. i, 669 (1840).

Widely distributed in the state but not very abundant; on sandy banks. Laramie Plains July 10, 1894 (No. 333); and B. C. Buffum, at Wheatland.

VIOLACEÆ.

Viola blanda, Willd. Hort. Berol. t, 24 (1806).

Quite rare, on mossy bank in the light spray of a little waterfall. Centennial Hills, June 9, 1895 (No. 1257).

Viola Canadensis, L. Sp. Pl. 936 (1753).

Abundant in thickets along streams.

Head of Pole Creek, May 25, 1894 (No. 44); Near Table Mountain, July 2, 1895 (No. 1406).

Viola canina adunca, Gray, Proc. Am. Acad. viii, 377.

Occurs less frequently than the following, leaves less crowded on the rootstock.

Pass Creek June 20, 1892, B. C. Buffum; Horse Creek, June 9, 1894 (No. 209).

Viola canina Muhlenbergii, Traut. Act. Hort. Petrop. v, 28.

Widely distributed, rather abundant and variable.

Pole Creek June 2, 1894 (No. 146); Centennial Valley, June 9, 1895 (No. 1284).

Viola Nuttallii, Pursh, Fl. Am. i, 174 (1814).

The earliest and commonest Violet on the plains.

Laramie, May 19, 1894 (No. 37); May 23 1895 (No. 1229).

Viola palmata cucullata, Gray, Bot. Gaz. xi, 254 (1886).

Infrequent. Horse Creek, June 9, 1894 (No. 191).

Viola palustris, L. Sp. Pl. ii, 934 (1753).

A beautiful little plant observed but once, on a boggy bank. Pole Creek, June 2, 1894 (No. 140).

Viola præmorsa, Dougl. Lindl. Bot. Reg. i, 1254.

This must be very rare in the state and its occurrence here extends its range considerably. Observed both in 1894 and in 1895, only at the head of Pole Creek, May, (Nos. 43 and 1215).

CARYOPHYLLACEÆ.

Saponaria Vaccaria, L. Sp. Pl. 409 (1753).
    This is becoming a troublesome weed in some parts of the state. From the Big Horn Mountains, by B. C. Buffum in 1892; Wheatland, July 11, 1894 (No. 474).

Silene acaulis, L. Sp. Pl. ii. 603 (1762).
    Strictly alpine, Teton Mountains, August 21, 1894 (No. 973); Medicine Bow Mountains, August 22, 1895 (No. 1828).

Silene antirrhina, L. Sp. Pl. 419 (1753).
    Widely distributed and not rare. Platte River, July 14, 1894 (No. 493); Centennial Valley, August 17, 1895 (No. 1658).

Silene Douglasii multicaulis, Robinson, Contr. Gray, Herb. 144 (1893).
    Union Pass at 10,000 ft., August 13, 1894 (No. 1019).

Silene Douglasii viscosa, Robinson, l. c. 145.
    This is not quite typical but Dr. Robinson thinks it too near to be separated. Union Pass, August 11, 1894 (No. 845).

Silene Hallii, Watson, Proc. Am. Acad. xxi, 446.
    On grassy, open slopes at high elevations; LaPlata Mines, August 21, 1895 (No. 1829).

Agrostemma Githago, L. Sp. Pl. 435 (1753).
    As yet very rare in the state. Collected at Sugg's Road by B. C. Buffum August, 1891.

Cerastium alpinum Behringianum, Regel. Ost. Sib. i, 435.
    Not common even in the mountains; Union Peak, August 13, 1894 (No. 1013).

Cerastium arvense, L. Sp. Pl. 438 (1753).
    Very abundant in early summer in the Laramie Hills. Pole Creek, June 2, 1894 (No. 138); near City Springs, June 21, 1891, B. C. Buffum.

Cerastium arvense latifolium, Fenzl. Ledeb. Fl. Ross. i, 412.
    Frequent on rocky hills and ledges in the mountains; Laramie Hills, May 24, 1894 (No. 41).

Cerastium arvense maximum, Hollick and Britton, Bull. Torr. Club, xiv, 45.
    Apparently rare; near the creek bank on lower Pole Creek, July 1, 1895 (No. 1380).

Cerastium nutans, Raf. Prec. Dec. 36 (1814). *C. longipedunculatum* Muhl.

Collected only in the north-western part of the state; Bacon Creek, August 16, 1894 (No. 924).

Stellaria borealis, Bigel. Fl. Bost. Ed. 2. 182 (1824). *Alsine borealis,* (Bigel.) Britton.

Infrequent; Centennial Valley, August 18, 1895 (No. 1738).

Stellaria longifolia, Muhl. Willd. Enum. 479. *Alsine longifolia,* (Muhl) Britton.

Abundant in wet places along streams; near Table Mountain, July 2, 1895 (No. 1417).

Stellaria longifolia laeta, T. & G. Bibl. Index. 112.

Rare; observed but once, LaPlata Mines, August 21, 1895 (No. 1774).

Stellaria longipes, Goldie, Edinb. Phil. Journ. vi, 327. *Alsine longipes,* (Goldie) Coville.

The commonest of the Chickweeds in Wyoming. Horse Creek, July 11, 1891, B. C. Buffum; Laramie, June 30, 1894 (No. 286). Lander, August 4, 1894 (No. 713).

Stellaria umbellata, Turcz. Cat. Baic. 5.

Frequent in the Mountains at 9,000 to 11,000 ft.; Union Pass, August 13, 1894 (No. 992); LaPlata Mines, August 21, 1895 (No. 1809).

Arenaria congesta, Nutt. T. & G. Fl. i, 178.

In open and in grassy places on hills and in the mountains everywhere; immensely abundant; Little Bald Mountain, August 15, 1892, B. C. Buffum; Laramie Hills, July 7, 1894 (No. 357).

Arenaria congesta subcongesta, Watson, Bot. Cal. i, 60.

Infrequent; only at high altitudes; on the Grand Teton, at 10,000 ft., August 21, 1894 (No. 1059).

Arenaria Fendleri, Gray, Pl. Fendl. 13.

This, like *A. congesta,* is found everywhere in the hills and mountains, in dry open rather than shaded ground. Laramie Hills, July 7, 1894 (No. 353).

Arenaria Hookeri, Nutt. T. & G. Fl. N. A. i, 178 (1838).

This is of very frequent occurrence both on the plains and in the mountains. Somewhat variable in general appearance; that on the plains short and forming large mats; that in the hills growing in the Pine needles in the shade less, cæspitose and of ranker growth. Laramie, June 20, 1894 (No. 225); Wind River Mountains, August 11, 1894 (No. 856); Laramie Peak, August 7, 1895 (No. 1595).

Arenaria Sajanensis, Willd. Schlecht. Berl. Mag. Natf. 200 (1816).

This plant collected at almost the opposite extremes of the state seems to confine itself to the naked alpine summits. Union Peak, August 13, 1894 (No. 1009); La Plata Mines, August 22, 1895 (No. 1826).

Arenaria sp. (See appendix).

Tissa sparsiflora, Greene, Erythea, iii, (1895).

This recently described species seems to have a very circumscribed range. It was first observed in the autumn of 1894, when it was collected by the writer in a wet meadow, some seven miles from Laramie, in a soil strong with alkali, receiving seepage water from an irrigation ditch. A very rank growth, resulting in long, lax and sparsely flowered stems had been attained. Observations upon it in 1895 in the same and other localities show that, under normal conditions, it grows to only 3-8 inches in height; that it is nearly erect, but freely branched from the base. This shortening of the stems shows the flowers to be numerous in proportion to the size of the plant and makes the specific name hardly characteristic. In the original description the observation is made that it is the first Tissa reported from the interior of the continent. Observed only about Laramie and in low alkali ground. Seven Mile Lake, October 15, 1894 (No. 1158); Laramie, September 3, 1895 (No. 1868).

PORTULACACEÆ.

Portulaca oleracea, L. Sp. Pl. 445 (1753).

Becoming introduced in some localities; from Sheridan Expt. Farm, by the superintendert, J. F. Lewis, September 1895.

Calandrinia pygmæa, Gray, Proc. Am. Acad. viii, 623.

A beautiful little plant of alpine habitat. Union Peak, August 13, 1894 (No. 1015); La Plata Mines, August 20, 1895 (No. 1778).

Claytonia Caroliniana sessilifolia, Torr. Pac. R. Rep. iv, 70 (1856).

Common on hillsides in rich, damp soil. Laramie Hills, May 12, 1894 (No. 27); Pole Creek, May 18, 1895 (No. 1219).

Claytonia sp.?
This may be a reduced alpine form of the preceding. The whole plant is small, 1-2 inches high, raceme reduced to one or two flowers and the leaves more acutely lanceolate. Collected on the shores of a lake at 9,000 ft. on the Grand Teton, August 21, 1894 (No. 1061).

Claytonia Chamissonis, Esch. & Spreng. Syst. i, 790 (1825).
Not of frequent occurrence, but sometimes growing in the greatest profusion on the rocky beds of slow-flowing brooklets; Sybille Creek, July 8, 1894 (No. 309); Pole Creek, June 30, 1895 (No. 1337).

Lewisia rediviva, Pursh, Fl. ii, 368 (1814).
Comparatively rare, but occurring occasionally in profusion among the sage brush on the plains, and sometimes in the pine needles of rather open woods in the foothills. Sweetwater River, June 22, 1891, D. McLaren; Garfield Peak, July 29, 1894 (No. 679); Cummins, July 27, 1895 (No. 1545).

HYPERICACEÆ.

Hypericum Scouleri, Hook. Fl. Bor. Am. i, 111 (1830). *H. formosum Scouleri,* (Hook.) Coult.
Quite rare; in thickets along streams. Sybille Creek, July 8, 1894 (No. 341).

MALVACEÆ.

Sidalcea candida, Gray, Pl. Fendl. 20 and 24.
Frequent and abundant in thickets along streams in the mountains at 8,000-9,000 ft. Centennial Valley, September 8, 1891, B. C. Buffum; Cummins, July 31, 1895 (No. 1489).

Sidalcea malvæflora, Gray, Pl. Wright. i, 16 (1852).
Habitat and localities similar to those of the preceding; possibly of more frequent occurrence, the two species sometimes growing together. Saratoga, July 2, 1893, J. D. Parker; Cummins, July 28, 1895 (No. 1463).

Malvastrum coccineum, Gray, Mem. Am. Acad. iv, 21 (1848).
A common weed on the plains, in fields and fence corners. University Campus, June 22, 1894 (No. 280).

Sphæralcea acerifolia, Nutt. T. & G. Fl. N. A. i, 228 (1838).

Not abundant but widely distributed; in open woods at 8,000-9,000 ft. Union Pass, August 10, 1894 (No. 873); Centennial Valley, August 17, 1895 (No. 1727).

Sphæralcea Munroana, Spach. Proc. Am. Acad. xxii, 292 (1887).

Laramie, September 1, 1893; Sheridan Experiment Farm, September 1895.

LINACEÆ.

Linum Kingii, Watson, King Rep. v, 49 (1871).

A few specimens of this species were secured by Prof Buffum in 1892, but without data; probably collected late in June, near Elk Mountain.

Linum perenne, L. Sp. Pl. 277 (1753). *L. Lewisii,* Pursh.

Remarkably abundant and luxuriant throughout the state. Found on dry, rocky ridges as well as on rich hillsides and valleys. The valley of Bacon Creek in August presents the appearance of a flax field. Laramie, June 12, 1894 (No. 241); Union Pass and Bacon Creek, August 1894 (No. 866).

Linum rigidum, Pursh, Fl. Am. Sept. 210 (1814).

Common in the eastern part of the state. Inyan Kara Divide, August 29, 1892, B. C. Buffum; Wheatland, July 9, 1894 (No. 384).

GERANIACEÆ.

Geranium cæspitosum, James, Long's Exped. ii, 3.

Very frequent in the Laramie Hills, growing in scattered clumps on rocky ridges. By B. C. Buffum in 1892; Telephone Canon, June 15, 1894 (No. 233).

Geranium Fremonti, Torr. Gray, Pl. Fendl. 26.

Rare, observed but once; Union Pass, August 11, 1894 (No. 824).

Geranium Richardsoni, Fisch. Mey. Ind. Sem. Petr. iv, 37.

Frequent in the south-eastern part of the state at least; along streams and in damp thickets. Pole Creek, June 2, 1894 (No. 132); near Table Mountain, July 1, 1895 (No. 1403).

Oxalis stricta, L. Sp. Pl. 281 (1753).

This is the only species so far found in the state and this but once. Whalen Canon, July 18, 1894 (No. 522).

CELASTRACEÆ.

Pachystima Myrsinites, Raf. Am. Month. Mag. 176 (1819).
Occurring on the sides of wooded mountains. Tetons, August 22, 1894 (No. 977), at 7,500 ft.

RHAMNACEÆ.

Ceanothus Fendleri, Gray Pl. Fendl. 29 (1849).
Very rare, a single clump of it in an open valley, Laramie Peak, August 8, 1895 (No. 1637.)

Ceanothus velutinus, Dougl. Hook. Fl. Bor. Am. i, 125 (1830).
Presumably throughout the state; dry canon-sides; Beaver Creek by B. C. Buffum, July 17, 1892; Tetons, August 21, 1894 (No. 948); Cummins, July 31, 1895 (No. 1542).

VITACEÆ.

Vitis riparia, Michx. Fl. ii, 231. *Wild Grape.*
This was collected in fruit not yet ripe, in one locality only; on the banks of the Platte River, Fairbanks, July 14, 1894 (No. 468).

Ampelopsis quinquefolia, Michx. Fl. Bor. Am. i, 160 (1803) *Parthenocissus quinquefolia,* (L.) Planch.
Quite rare in the state, possibly not found in the more elevated districts at all. Hartville, July 16, 1894 (No. 554).

SAPINDACEÆ.

Acer glabrum, Torr. Ann. Lyc. N. Y. ii, 172 (1826).
A common shrub on rocky hillsides and in the canons. Telephone Canon, May 23, 1894 (No. 57); Laramie Hills May 25, 1895 (No. 1236). *Maple.*

Negundo aceroides, Moench. Meth. 334 (1794). *Acer Negundo,* L.
Occurring occasionally along streams. Introduced at Laramie for shade and decorative purposes. June 1, 1894 (No. 183); Big Muddy Creek, July 26, 1894 (No. 611). *Box Elder.*

ANACARDIACEÆ.

Rhus toxicodendron, L. Sp. Pl. (1753). *Rhus radicans,* L.
Ours is the low erect form. Among rocks in canons at 5,000-6,000 ft. Table Mountain, June 2, 1894 (No. 154); Hartville, July

16, 1894 (No. 557); noted also at Laramie Peak, growing in profusion in a rocky canon. *Poison Ivy.*

Rhus trilobata, Nutt. T. & G. Fl. i, 219 (1838).

A low, spreading shrub, frequently almost covering both sides and summits of the low rounded hills in the Laramie range. Table Mountain, June 2, 1894 (No. 159); Blue Grass Hills, July 8, 1894 (No. 322); Laramie Peak, August 7, 1895 (No. 1477).

LEGUMINOSÆ.

Thermopsis montana, Nutt. T. & G. Fl. i, 388 (1838).

Not common, occurring on sandy creek banks. Laramie River, fl. June 15, fr. August 19, 1891, B. C. Buffum.

Thermopsis rhombifolia, Richards. App. Frank. Journ. 13 (1823).

Abundant in sandy ravines and valleys in the hills, the great patches of yellow standing out in sharp contrast to the green grass and white rocks. Table Mountain, June 2, 1894 (No. 122); Laramie Hills, June 5, 1895 (No. 1240).

Lupinus argenteus, Pursh, Fl. 468 (1814).

A widely distributed and common species of this well represented and beautiful genus. Apparently at higher altitudes than the var. following. Chugwater Creek, July 7, 1894 (No. 301); Meadow Creek, August 9, 1894 (No. 972); Laramie Peak, August 7, 1895 (No. 1584).

Lupinus argenteus decumbens, Wats. Proc. Am. Acad. xviii, 532 (1873).

On creek banks in the hills and plains. Pole Creek, June 2, 1894 (No. 104).

Lupinus aridus, Dougl. Lindl. Bot. Reg. xv, 1242 (1829).

On the plains of the Sweetwater River, by Geo. M. Cordiner, September 6, 1894 (No. 1206).

Lupinus cæspitosus, Nutt. T. & G. Fl. i, 379.

A delicate little plant almost alpine in its habitat; observed but once; Union Peak, August 13, 1894 (No. 996).

Lupinus laxiflorus, Dougl. Lindl. Bot. Reg. xiv, 1140 (1828).

Frequent and sometimes covering great stretches of the sandy plain with its characteristic color. Blue Grass Creek, July 8, 1894 (No. 360); Lusk, July 21, 1894 (No. 584); also west slope of Wind River Mountains, August 14, 1894.

Lupinus leucophyllus, Dougl. Lindl. Bot. Reg. xiii, 1124 (1828).

Our earliest Lupine, found in great profusion on moist hillsides among the sage brush and even in shaded localities. Laramie Hills, June 2, 1894 (No. 151).

Lupinus ornatus, Dougl. Lindl. Bot. Reg. xv, 1216 (1829).

Certainly deserving its name; abundant in the locality noted; Gros Ventre River, August 16, 1894 (No. 1098).

Lupinus parviflorus, Nutt. Hook. & Arn. Bot. Beechy, 336.

Common along streams; Sybille Creek, August 8, 1894 (No. 315); Table Mountain, July 1, 1895 (No. 1414).

Lupinus Plattensis, Wats. Proc. Am. Acad. xvii, 124 (1890).

This rare and beautiful plant was observed in two localities only. Mexican Mines, July 20, 1894 (No. 589); Pole Creek, near Table Mountain, July 1, 1895 (No. 1401).

Lupinus pusillus, Pursh, Fl. Am. Sept. 468 (1814).

Found only on the sand ridges and dunes occurring occasionally on the plains of eastern Wyoming. Platte River, July 14, 1894 (No. 490); noted also south of Lusk.

Lupinus Sitgreavesii, Wats. Proc. Am. Acad. xiii, 527.

Occurs only at comparatively high altitudes in wooded mountains. Union Pass, August 12, 1894 (No. 896); frequent at 9,000 ft. and upward.

Medicago sativa, L. Sp. Pl. 778 (1753). *Lucerne, Alfalfa;* largely grown as a forage plant in the west; escaped from cultivation. Laramie, September 9, 1894 (No. 1136).

Melilotus alba, Lam. Encycl. iv, 63 (1797). *Sweet Clover.*

Persisting in fallow or abandoned fields. Laramie, October 2, 1894 (No. 1154.)

Melilotus officinalis, (L.) Lam. Fl. France, ii, 594 (1778).

Introduced and then persisting in abandoned areas for a number of years, possibly indefinitely. Laramie, June 23, 1895 (No. 1422).

Psoralea argophylla, Pursh, Fl. Am. Sept. 475 (1814).

Noted a number of times in eastern Wyoming; Platte River, July 14, 1894 (No. 497); from Inyan Kara Divide by B. C. Buffum, 1892, and from Sheridan by J. F. Lewis, 1895.

Psoralea lanceolata, Pursh, l. c.

Frequent on the dry foothills along the Platte River. Fairbanks, July 12, 1894 (No. 430); Orin Junction, August 14, 1892, B. C. Buffum.

Psoralea tenuiflora, Pursh, l. c.

Our commonest Psoralea, very abundant along the Platte and its tributaries. Laramie River, July 10, 1894 (No. 368); Blue Grass Creek, July 8, 1894 (No. 306).

Trifolium dasyphyllum, T. & G. Fl. N. A. i, 315.

This fine cæspitose species almost clothes some of the otherwise naked rocky ledges in the Laramie foothills. Some specimens secured at Laramie Peak are quite erect with longer and less pubescent leaves. Laramie, May 25, 1894 (No. 68); June 5, 1895 (No. 1243).

Trifolium gymnocarpon, Nutt. T. & G. Fl. i, 320.

A rare and inconspicuous little plant, blossoming in late spring and shortly disappearing. Laramie, June 9, 1894 (No. 216); Experiment Farm, May 23, 1895 (No. 1230).

Trifolium longipes, Nutt. T. & G. Fl. i, 314 and 691.

Frequent in the mountains at 8,000-9,000 ft. Specimens from Saratoga and Bald Mountain; also Union Pass, August 13; 1895 (No. 1025).

Trifolium longipes reflexum, n. var.

This has the general habit of *T. longipes* but the flowers are at length quite reflexed; calyx lobes shorter and less villous. On the banks of Wind River at the foot of Union Pass, August 9, 1894 (No. 918).

Trifolium Parryi, Gray, Am. Journ. Sci. ii, 33.

Rare, in open spruce woods, Medicine Bow Mountains, August 21, 1895 (No. 1764).

Trifolium pratense, L. Sp. Pl. 768 (1753). *Red Clover.*

Trifolium repens, L. Sp. Pl. 767 (1753).

This and the preceding becoming naturalized along irrigation ditches, in the streets and elsewhere. Laramie, September 15, 1894. *White Clover.*

Amorpha fruiticosa, L. Sp. Pl. 713 (1753).

Frequent on the banks of the Platte in the eastern part of the state. Fairbanks, July 13, 1894 (No. 438).

Dalea aurea, Nutt. Fras. Cat. (1813). *Parosela aurea*, (Nutt.) Britton.
Very rare; on the plains of the Platte; Fairbanks, July 11, 1894 (No. 390).

Petalostemon candidus, Michx. Fl. Bor. Am. ii, 49 (1803). *Kuhnistera candida*, (Willd.) Kuntze.
Occasional, on the dry hills and plains bordering on the Platte. Orin Junction, August 14, 1891, B. C. Buffum.

Petalostemon multiflorus, Nutt. Journ. Phil. Acad. vii, 92 (1834).
This seems to be a much-named plant and illustrates nicely the stability that our nomenclature is acquiring. The new Check List gives *Kuhnistera multiflora*, (Nutt.) Heller, and, if I understand Mr. Rydberg rightly, his new name, *Kuhnistera, candida multiflora*, (Nutt.), Contrib. Nat'l Herb. iii, 3. (1895), is also the same. From the specimens at hand I am inclined to think that Mr. Rydberg is right in reducing the form to a variety. On the other hand his variety *occidentalis* does not seem to differ in any important respect from his *multiflora*, judging by his descriptions. Some specimens at hand will fall nicely under either.
Very frequent in the Laramie Hills and the foothills bordering on the Platte. July 9, 1894 (No. 330); Inyan Kara Divide by B. C. Buffum.

Petalostemon violaceus, Michx. Fl. Bor. Am. ii, 50 (1803). *Kuhnistera purpurea*, (Vent.) MacM.
Habitat and localities much the same as for the preceding. Orin Junction, July 14, 1891, B. C. Buffum; Platte Hills, July 9, 1894 (No. 331).

Astragalus adsurgens, Pall. Astrag. 40. t, 31 (1800). *A. Laxmanni*, Jacq.
Remarkably abundant in the south-eastern part of the state, occupying dry, stony or gravelly ridges on the plains or in the foothills.
Wallace Creek, July 29, 1894 (No. 646); Cummins, July 30, 1895 (No. 1514); at Laramie at various times.

Astragalus alpinus, L. Sp. Pl. 760 (1753).
Probably frequent in the higher mountain valleys; Pole Creek, June 3, 1894 (No. 174); Union Pass, August 11, 1894 (No. 840). Unusually large specimens with leaves varying from elliptical to obcordate were obtained on Union Peak, August 13, 1894 (No. 993).

Astragalus bisulcatus, Gray, Pac. R. Rep. xii, 42 (1860.)
   A very common species on the plains and in the foothills; Laramie, June 19, 1894 (No. 266); Garfield Peak, July 29, 1895 (No. 682); other numbers are 1316 and 1435.

Astragalus Canadensis, L. Sp. Pl. 757 (1753). *A. Carolinianus,* L.
   Possibly confined to the eastern part of the state.
   Lusk, July 21, 1894 (No. 582); Laramie Peak, August 7, 1895 (No. 1597).

Astragalus caryocarpus, Ker. Bot. Reg. t, 176 (1816). *A. crassicarpus,* Nutt.
   Infrequent; on the Laramie plains and on the east slopes of the Laramie range. Pole Creek, June 2, 1894 (No. 162); Laramie, June 25, 1894 (No. 201).

Astragalus convalarius, Greene.
   I am unable to cite the publication of this; the name was communicated by Prof. Sheldon; probably rare; Union Pass, August 10, 1894 (Nos. 743 and 869).

Astragalus Drummondii, Dougl. Hook. Fl. Bor. Am. i, 153 (1833).
   Frequent in the foothills throughout the state; Pole Creek, June 2, 1894 (No. 86); Upper Wind River, August 10, 1894 (No. 763).

Astragalus flexuosus, Dougl., in Don. Gen. Syst. Gard. and Bot. ii, 256 (1832).
   On wet, fertile creek banks; Chugwater, July 7, 1894 (No. 422); Pole Creek, July 1, 1895 (No. 1352).

Astragalus frigidus Americanus, Watson, Index, i, 193.
   Very rare; South Fork, Crazy Woman Creek, August 7, 1892, B. C. Buffum.

Astragalus giganteus, Sheld. Bull. Minn. Geol. and Nat. Hist. Surv. ix. 65 (894).
   Infrequent; Bald Mountain, August 15, 1892; Green River, August 26, 1894 (No. 1047).

Astragalus hypoglottis, L. Mant. ii, 274 (1771).
   Frequent and abundant in wet meadows and along streams in our whole range. Saratoga, July 6, 1891; Meadow Creek, August 9, 1894 (No. 815). Fine specimens with ohroleucus flowers were obtained at Meadow Creek, August 9, 1894 (No. 775).

Astragalus Kentrophyta, Gray, Proc. Acad. Philad. 60 (1863).
   Probably belonging to the Pacific slopes alone; on the banks of the Gros Ventre River, August 16, 1894 (No. 1077).

Astragalus junceus, Gray, Proc. Am. Acad. vi, 230.
   Infrequent; on steep, dry stony hillsides; Saratoga, June 23, 1893; Gros Ventre River, August 17, 1894 (No. 1086).

Astragalus lonchocarpus, Torr. Pac. R. Rep. iv, 80.
   What appears to be of this species was collected on Snake River, May 29, 1892, by Fred McCoullough.

Astragalus Missouriensis, Nutt. Gen. ii, 99 (1818).
   Abundant on the Laramie plains but in the foothills giving place to *A. Shortianus*. Laramie, May 22, 1894 (No. 52); also in 1895 (No. 1227).

Astragalus Mortoni, Nutt. Journ. Acad. Phila. vii, 19 (1834).
   On the Pacific slope only, infrequent; Gros Ventre River, August 16, 1894 (No. 1080).

Astragalus oroboides Americanus, Gray, Proc. Am. Acad. vi, 205 (1864). *A. oroboides*, Hornem.
   Infrequent; Bacon Creek, August 15, 1894 (No. 917).

Astragalus pectinatus, Dougl. in Don. Gen. Syst. Gard. and Bot. ii, 257 (1832).
   Frequent and abundant on the Laramie Plains; June 12, 1894 (No. 217).
   Some forms of it seem to approach *A. Grayi* so closely as to make it difficult to know where to place them, such as my number 1304 from Centennial Valley, June 9, 1895.

Astragalus Purshii, Dougl. Hook. Fl. Bor. Am. i, 152 (1834).
   Of frequent occurrence, but the plants few and scattering.
   On the Laramie Plains, May 23, 1894 (No. 53); May, 1895 (No. 1228).

Astragalus sericoleucus, Gray, Am. Jour. Sci. II, xxxiii, 410 (1862).
   Frequent on the plains and in the foothills, where it clothes the otherwise naked ground as with a purple carpet.
   Laramie, May 18, 1894 (No. 38).

Astragalus Shortianus, Nutt. T. & G. Fl. N. A. i, 331 (1838).
   On gravelly, stony hillsides; frequent in the Laramie range.
   Telephone Canon, May 23, 1894 (No. 54); Centennial Valley, June 9, 1895 (No. 1286).

Astragalus spatulatus, Sheld. Bull. Minn. Geol. & Nat. Hist. Surv. ix, 19 (1894).

Very abundant both on the plains and in the foothills.

Laramie Hills, May 16, 1894 (No. 31); Centennial Valley, June 9, 1895 (No. 1301); also a white flowered variety from the latter place.

Astragalus tenellus, Pursh, Fl. Am. Sept. 473 (1814).

Seemingly throughout the state, in dry, sandy soil.

Laramie, June 19, 1894 (No. 267); Dubois, August 9, 1894 (No. 751); Cummins, July 26, 1895 (No. 1432).

Astragalus. (For other numbers see appendix).

Oxytropis deflexa, D C. Prodr. ii ,280.

Infrequent, in mountain Meadows, at 7,000-8,000 ft.; Union Pass, August 11, 1894 (No. 825); Laramie Peak, August 8, 1895 (No. 1530).

Oxytropis Lamberti, Pursh, Fl. Am. Sept. 740 (1814). *Spiesia Lamberti,* (Pursh) Kuntze.

The typical form is not frequent. As far as my observation goes, it is confined to the eastern part of the state. By the typical form is meant the purple-flowered form described in our manuals under this name. Table Mountain, July 1, 1895 (No. 1320).

Oxytropis Lamberti sericea, Nutt. T. & G. i, 339 (1838). *Spiesia Lamberti sericea,* (Nutt.) Rydberg.

Very rare; Laramie, 1893. Flowers violet.

Oxytropis (Spiesia) Lamberti ochroleuca, n. var.

Stout, grayish, with a close pubescence throughout; very many short stems from the large perennial root, each of which bears one to several long scape like peduncles; flowers yellowish-white with sometimes a purple spot on the keel petals.

This is the frequently mentioned ochroleucus flowered form of this species, but it certainly, with us at least, forms a good and constant variety, if not species. It can always be distinguished by its stouter habit throughout, by the densely lanate scale-like stipules, by the shorter but thicker spike as well as by the very numerous and crowded stems and scapes. Exceedingly abandant; Pole Creek, June 2, 1894 (No. 119); Laramie, June 9, 1895 (No. 1302). *Loco.*

Oxytropis monticola, Gray, Proc. Am. Acad. xx, 6.
  Frequent in the vicinity of Laramie; May 23, 1894 (No. 60); June 9, 1895 (No. 1294).

Oxytropis multiceps, Nutt. T. & G. Fl. i, 341 (1838). *Spiesia multiceps,* (Nutt.) Kuntze.
  Very rare; on a hilltop of disintegrated granite, near Horse Creek, June 9, 1894 (No. 214).

Oxytropis splendens, Dougl. Hook. Fl. Bor. Am. i, 147 (1833). *Spiesia splendens,* (Dougl.) Kuntze.
  This truly splendid species is not rare in the grassy valleys at about 7,000-8,000 ft. and sometimes at lower altitudes. Table Mountain, July 2, 1895 (No. 1391); Cummins July 29, 1895.

Oxytropis sp. ? (Nos. 669 and 928, see appendix).

Glyceria lepidota, Pursh, Fl. Am. Sept. 480 (1814).
  Frequent near the Platte on the banks of small ditches and ravines. Willow Creek, July 22, 1894 (No. 627); Cheyenne, August 29, 1893.

Hedysarum boreale, Nutt. Gen. ii, 110 (1818). *H. Americanum,* (Michx.) Britton.
  Probably infrequent except northward; Union Pass, August 10, 1894 (No. 877).

Hedysarum Mackenzii, Richards, App. Frank. Journ. 17 (1823). *H. Americanum Mackenzii,* (Richards.) Britton.
  Throughout the state; Crazy Woman Creek, August 7, 1892; Dubois, August 9, 1894 (No. 752); Snake River, August 19, 1894 (No. 1087).

Vicia Americana, Muhl. Willd. Sp. Pl. ii, 1096 (1803).
  Of frequent occurrence in thickets on stream banks. Cummins, July 29, 1895, two forms of it, (Nos. 1450 and 1478).

Vicia Americana truncata, Brewer, Bot. Cal. i, 158 (1876).
  More frequent than the species; Wheatland, August 1892, B. C. Buffum; Table Mountain, June 30, 1895 (No. 1404).

Vicia linearis, (Nutt.) Greene, Fl. Fran. 3 (1891).
  Especially abundant; on sandy plains and creek valleys everywhere.
  Pole Creek, June 2, 1894 (No. 92); Laramie, June 1, 1894 (No. 172); Meadow Creek, August 9, 1894 (No. 776).

Vicia cracca, L. Sp. Pl. 735 (1753).

A chance introduction on the Laramie Experiment Farm, September 1895.

Lathyrus ornatus, Nutt. T. & G. Fl. N. A. i, 277 (1838).

Members of this genus are either very rare or the right localities have not yet been visited. A specimen in the herbarium from Cheyenne, by Miss Helen Furniss, June 1892.

ROSACEÆ.

Prunus Americana, Marsh, Arb. Am. 111 (1785).

Rare in the southeastern part of the state and probably infrequent everywhere except in the lower altitudes of the northeastern part, where it is reported very plentiful.

Fairbanks, July 18, 1894 (No. 572). *Wild Plum.*

Prunus demissa, Walp. Rep. ii, 10 (1843).

This and the succeeding species have so much in common that it is difficult to separate them. It may be that none of the specimens are *P. Virginiana,* but rather only forms of *P. demissa.* Common on dry creek banks in the hills.

Telephone Canon, June 15, 1894 (No. 230); Table Mountain, June 30, 1895 (No. 1402). *Wild Cherry, Choke Cherry.*

Prunus Virginiana, L. Sp. Pl. 473 (1753).

Teton Mountains, August 21, 1894 (No. 943), and by B. C. Buffum, 1892.

Spirea arbuscula, Greene, Erythea, iii, 63 (1895).

Infrequent and, I think, found only at high elevations, 8,000 ft. and upward.

Teton Mountains, August 21, 1894 (No. 941).

Spirea discolor dumosa, Watson, Pursh Fl. 342 (1814).

A handsome shrub, common on rocky ledges at 8,000 ft. and upward.

Casper Mountain, July 26, 1894 (No. 607); Garfield Peak, July 29, 1894 (No. 657); also observed at Laramie Peak and in the Medicine Bow Mountains in 1895.

Spirea lucida, Dougl.

Very rare in the state.

Teton Mountains, August 21, 1894 (No. 949).

Physocarpus Torreyi, Maxim. *Neillia Torreyi,* Watson, Proc. Am. Acad. xi, 136.

Common in the hills and mountains.

Table Mountain, June 2, 1894 (No. 116); Platte River, July 14, 1894 (No. 490).

Rubus Nutkanus, Mocino, Lindl. Bot. Reg. t. 1368 (1830).

Not common in the parts of the state collected.

Teton Mountains, August 21, 1894, and Centennial Hills, August 18, 1895 (No. 1676). *Thimble-berry.*

Rubus strigosus, Michx. Fl. i, 297 (1803).

The *Red Raspberry,* growing in the greatest profusion on rocky and partially wooded hillsides, especially on ground once burned over. Immensely productive.

Union Peak, August 13, 1894 (No. 997). Cummins, July 31, 1895 (No. 1476).

Purshia tridentata,, D C. Trans. Linn. Soc. 12, 157.

A scragly prostrate shrub, common on low hilltops and hillsides.

Pole Creek, June 2, 1894 (No. 82); Wallace Creek, July 29, 1894 (No. 676); Centennial Valley, June 9, 1895 (No. 1270).

Cercocarpus parvifolius, H. & A. Bot. Beechey, 337 (1841).

This may be called the most characteristic shrub of stony foothills.

Laramie Hills, June 15, 1894 (No. 237); Platte Hills, July 14, 1894 (No. 462). *Mountain Mahogany.*

Geum macrophyllum, Willd. Enum. i, 557 (1809).

Mr. Rydberg* suggests the probability that this and the following are varieties of the same species. The specimens before me, however, seem perfectly distinct. Near streams.

Bacon Creek, August 15, 1894 (No. 920); Cummins, July 30, 1895 (No. 1576).

Geum strictum, Ait. Hort. Kew. ii, 217 (1789).

Same habitat as the preceding.

Sybille Creek, July 8, 1894 (No. 407); Cummins, July 30, 1895 (No. 1517).

Geum triflorum, Pursh, Fl. 736 (1814).

This fine species is common in wet valleys at 7,000-9,000 ft. Our manuals fail to note that the style is jointed in the middle in young

---
*Contrib. Nat. Herb. iii, 3, 157 (1895).

specimens, but as the apical portion is early deciduous it appears perfectly straight in maturer blossoms.

Union Pass, August 11, 1894 (No. 829); Saw Mill Creek, May 25, 1895 (No. 1258).

Fragaria vesca Americana, Porter, Bull. Torr. Bot. Club, xvii, 15 (1890).

In mountain meadows and valleys, in parks and on wet hillsides everywhere. Sometimes fruiting abundantly, the berries small but sweet.

Horse Creek, June 9, 1894 (No. 207). *Strawberry*.

Fragaria Virginiana Illinœnsis, Prince, Gray Man Ed. 5, 155 (1867).

This species is rare, at least about Laramie, but specimens collected by A. H. Danielson near Jelm Mountain, May, 1895, seem to be of this form (No. 1209).

Potentilla anserina, L. Sp. Pl. 495 (1753).

Common in wet soils, especially near slightly alkali marshes. Laramie all summer.

Collected on Green River, August 26, 1894 (No. 1039).

Potentilla arguta, Pursh, Fl. Am. Sept. 736 (1814).

Forming immense yellow patches among the rocks in the hills. A fine plant.

Pole Creek Hills, June 2, 1894 (No. 95); June 30, 1895 (No. 1351).

Potentilla dissecta, Pursh, Fl. Am. Sept. (1814).

This fine species is probably confined to high elevations.

Teton Mountains, August 22, 1894 (No. 970); LaPlata Mines, August 20, 1895 (No. 1773).

Potentilla fruticosa, L. Sp. Pl. 495 (1753).

Common on the banks of mountain streams.

Wind River, August 9, 1894 (No. 748); Little Sandy, August 30, 1894 (No. 1127).

Potentilla glandulosa, Lindl. Bot. Reg. six, t. 1583 (1835).

Occasional along streams in sandy loam.

Union Pass, August 12, 1894 (No. 867); Cummins, July 30, 1895 (No. 1493).

Potentilla gracilis, Dougl. Hook. Bot. Mag. (1830).

This is a most polymorphous species, the forms of it differing strikingly as to tomentum, hirsuteness, leaf margin, etc. It seems probable that a careful examination will show at least some varieties which ought to be separated from the species. In wet meadows and along mountain streams.

Laramie River, July 9, 1894 (No. 325); Wallace Creek, July 29, 1894 (No. 665); Wind River Mountains, August 13, 1894 (No. 931); Table Mountain, June 30, 1895 (No. 1347); La Plata Mines, August 21, 1895 (No. 1821).

Potentilla gracilis fastigiata, Watson, Proc. Am. Acad. viii, 557 (1873).

Union Pass, August 13, 1894 (No. 990); Centennial Valley, August 25, 1895 (No. 1858).

Potentilla gracilis flabelliformis, Torr. & Gray, Fl. (1838).

A rare and beautiful plant, the deeply pinnatifid leaflets densely white tomentose on the lower surface.

Meadow Creek, August 9, 1894 (No. 786).

Potentilla Hippiana, Lehm. Nov. Stirp. Pug. ii, 197 (1830).

Very common on rocky slopes and hills.

Laramie Hills, July 7, 1894 (No. 410); Garfield Peak, July 29, (No. 651); Table Mountain, June 29, 1895 (No. 1368).

Potentilla Hippiana pulcherrima, Watson, Proc. Am. Acad. viii, 555 (1873). *Potentilla pulcherrima*, Lehm.

This is not common, and seems to have affinities with *P. gracilis*, *P. Pennsylvanica* and *P. Hippiana*.

La Plata Mines, August 22, 1895 (No. 1789).

Potentilla humifusa, Nutt. Gen. i, 330 (1818).

Common early in the year on gravelly hillsides.

Laramie Hills, June 2, 1894 (No. 99); May 18, 1895 (No. 1216).

Potentilla Norvegica, L. Sp. Pl. 499 (1753). *P. Monspeliensis*, L.

Apparently throughout the state.

Platte River and Willow Creek, July 1894 (Nos. 502 and 565); Laramie, September 16, 1894 (No. 1143); Cummins, July 30, 1895 (No. 1474).

Potentilla Pennsylvanica, L. Mant. 76 (1767).

Observed but once.

Bessemer, on the Platte River, July 26, 1894 (No. 612).

Potentilla Pennsylvanica strigosa, Pursh, Fl. Am. Sept. 356 (1814).

In rather dry valleys; more frequent than the species.

Whalen Canon, July 18, 1894 (No. 525); Cummins, July 30, 1895 (No. 1545).

Potentilla Plattensis, T. & G. Fl. i, 439 (1838).

On sandy flats bordering on streams.

Horse Creek, June 9, 1894 (No. 206); Laramie River, June 19, 1894 (No. 265).

Potentilla pinnatisecta, n. sp.

This form should not have been suppressed. In Watson's King's Report it appears as *P. diversifolia pinnatisecta*. The points upon which he based the variety are well taken but not strongly enough emphasized. The following, it seems to me, justify the separation of this form from *P. Plattensis* as a distinct species. *P. Plattensis* has been repeatedly observed but only on bottom lands near streams and, I think, thus far, with one exception,* only on streams tributary to the Platte. *P. pinnatisecta* is strictly alpine. The specimens before me were collected on the naked summits of the Medicine Bow Mountains, above timber line at about 11,500 feet, growing among the rocks in the most bleak and barren places.

The two differ markedly in habit as well as in habitat. *P. Plattensis* is decumbent at base, with branches diffuse or loosely spreading. *P. pinnatisecta*, while its branches are slightly decumbent at base, yet the plant as a whole may be spoken of as strictly erect. The cymes are more regular and more open. The leaves are longer, leaflets less crowded and inclined to be pedately parted rather than pinately, strictly cuneate at base. Leaves largely radical, those on the stem greatly reduced. Petioles and stems nearly glabrous with a brownish glaucus hue. Leaves glabrous on the upper surface, softly silky villous on the lower. The plant 5-8 inches high.

Potentilla supina, Michx. Fl. Bor. Am. i, 304 (1803). *P. paradoxa*, Nutt.

Not of frequent occurrence; Hartville, July 16, 1894 (No. 555); Cummins, July 28, 1895 (No. 1525).

Sibbaldia procumbens, L. Sp. Pl. 284 (1753).

Frequent at high altitudes, 10,000 ft. and upward. Union Peak, August 13, 1894 (No. 1010); La Plata Mines, August 23, 1895 (No. 1799).

---

*Bot. Death Valley Exped. Contrib. Natl. Herb. iv, 112 (1893).

Chamærhodos erecta, Bunge. Ledeb. Fl. Alt. i, 431.
  Observed but once but then in the greatest abundance, on a sand-bar of the Laramie River, June 18, 1894 (No. 261).

Agrimonia Eupatoria, L. Walt. Fl. Car. 131 (1788). *A. Striata,* Michx.
  Not at all common; Laramie Peak, on a wooded stream, August 9, 1895 (No. 1653).

Rosa acicularis, Lindl. Ros. Monog. 44, t. 8 (1820).
  Found most frequently at the foot of dry, clayey creek banks.
  Pole Creek, June 2, 1894 (No. 134); Centennial Valley, August 18, 1895 (No. 1697).

Rosa Arkansana, Porter, Syn. Fl. Col. 38 (1874).
  A fine species often met with in wet ravines in the mountains.
  Mexican Mines, July 20, 1894 (No. 567); Table Mountain, June 30, 1895 (No. 1336).

Rosa blanda, Ait. Hort. Kew. ii, 202 (1779).
  Rather common in canons and the adjacent hillsides.
  Whalen Canon, July 18, 1894 (No. 520); Table Mountain, June 30, 1895 (No. 1343).

Rosa Fendleri, Crepin, Prim. Monog. Ros. 452 (1874).
  Probably quite rare in the state.
  Boulder Creek, August 27, 1894 (No. 1101).

Rosa Sayi, Schwein. Keating, Long's Exped. Appx. 113.
  I note that recently this has been made a synonym for *R. acicularis,* but in Watson's Rev. Ros. N. A. 340 (1885), the two forms are kept distinct. If the two forms before me are correctly determined he is amply justified in doing so.
  Table Mountain, June 30, 1895 (No. 1643). Also by B. C. Buffum, June 6, 1892.

Rosa Woodsii, Lindl. Ros. Monog. 21 (1820).
  A small form, common, in the dry hills and sandy ravines, inclined to be prostrate.
  Laramie Hills, June 12, 1894 (No. 242).

Pyrus sambucifolia, Cham. & Schlecht. Linnæa, ii, (1827). *Sorbus sambucifolia,* Roem.
  Rather rare; Teton Mountains, August 22, 1894 (No. 982); also from the Centennial Valley, by B. C. Buffum.

Cratægus Douglasii, Lindl. Bot. Reg. 21, t, (1840).

Very scarce, only a few specimens observed in one locality. These are not typical, yet can hardly belong to the following species though approaching it more closely than any other.

Casper, July 25, 1894 (No. 666). *Hawthorn.*

Cratægus rivularis, Nutt. T. & G. Fl. i, 464 (1840).

Of rather frequent occurrence on some streams.

Gros Ventre River August 22, 1894 (No. 1066).

Amelanchier alnifolia, Nutt. Journ. Acad. Phila. viii, 22 (1834).

Very common in the foothills and along streams.

Telephone Canon, June 2, 1894 (No. 117).

Amelanchier pumila, Nutt. Roem. Syn. Monog. iii, 145.

Prof. E. L. Greene, of Catholic University, Washington, D. C., writes me that in 1893 I sent him some material as *A. alnifolia*, which, on examination, proves to be the long suppressed *A. pumila* of which he writes as follows: "It may be recognized by its being perfectly glabrous; by having much longer and narrower petals and by having its leaves serrate almost from base to apex, whereas *A. alnifolia* has its leaves serrate only from the middle upwards."

On examination of the material in our herbarium I find that our 1893 material answers to this description of *A. pumila* perfectly.

Laramie Hills, May 1893. *Service Berry.*

SAXIFRAGACEÆ.

Saxifraga bronchialis, L. Sp. Pl. i, 400 (1753).

Teton Mountains, at high elevations, August 21, 1894 (No. 976).

Saxifraga Jamesii, Torr. Ann. Lyc. N. Y. ii, 204.

This rare species was observed but once. Jackson's Hole, August 22, 1894 (No. 971).

Saxifraga integrifolia, Hook. Fl. Bor. Am. i, 249 (1834).

Grassy places in the mountains, but not at all common.

Bald Mountain, August 16, 1892, B. C. Buffum; Union Pass, August 10, 1894 (No. 930).

Saxifraga nivalis, L. Sp. Pl. i, 401 (1753).

This species is not strictly alpine if this idea be represented solely by altitude. Common, and blossoming in May at 7,000 ft., and successively throughout the season at higher and higher altitudes.

Wind River Mountains, July 20, 1892, B. C. Buffum; Pole Creek, June 2, 1894 (No. 127); La Plata Mines, August 22, 1895.

Saxifraga punctata, L. Sp. Pl. 401 (1753).
Frequent in wet, shaded ravines at high elevations.
Union Pass, August 12, 1894 (No. 843), also from Bald Mountain, by B. C. Buffum.

Saxifraga Virginiensis, Michx. Fl. Bor. Am. i, 269 (1803).
This is less common than *S. nivalis* which it closely resembles.
In the lower valleys of the state. Bald Mountain, August 15, 1892, by B. C. Buffum.

Tellima tenella, Walp. Rept. ii, 371 (1843).
Rare, on the sandy slopes of valleys.
Horse Creek, June 11, 1892; Bald Mountain, August 15, 1892, B. C. Buffum.

Mitella pentandra, Hook. Bot. Mag. lvi, t. 2933 (1829).
Frequent in damp ravines in the mountains.
Teton Mountains, August 21, 1894 (No. 944); Laramie Peak, August 8, 1895 (No. 1623).

Mitella trifida, Graham. var. **integripetala,** Rose.
This new variety is founded upon material collected by B. C. Buffum, and sent by myself to Dr. J. N. Rose, who named it and I suppose published it, but I am unable to cite the publication. It differs from the species mainly in its entire petals and its leaves, which are incisely dentate rather than crenate dentate. Certainly very rare.
Bald Mountain, August 15, 1892.

Heuchera parviflora, Nutt. T. & G. Fl. i, 581 (1838).
Very common on stony ridges and hillsides in the Laramie range.
Pole Creek, June 2, 1894 (No. 163); Table Mountain, June 28, 1895 (No. 1344); also observed in the Medicine Bow Mountains, August 1895.

Parnassia fimbriata, Banks, in Koenig & Sims, Ann. Bot. i, 391 (1805).
Along streams and in wet thickets at high altitudes.
Head of Green River, August 26, 1894 (No. 1034); La Plata Mines, August 22, 1895 (No. 1803).

Parnassia palustris, L. Sp. Pl. 273 (1753).

Quite rare and not readily distinguished from the next except by the cordate leaves.

Bald Mountain, August 16, 1892, B. C. Buffum.

Parnassia parviflora, D C. Prodr. i, 320.

Rather common on grassy stream banks almost in the water.

Fort Washakie, August 5, 1894 (No. 746).

Jamesia Americana, T. & G. Fl. i, 593 (1840).

A very handsome shrub when in blossom. Rare; in rocky canons.

From Saratoga by B. C. Buffum, June 1892; Table Mountain, July 1, 1895 (No. 1418).

Ribes aureum, Pursh, Fl. 164 (1814).

Common in the canons of the Platte, where the black fruited form seems to predominate. Introduced in Laramie as an ornamental shrub.

Fairbanks, July 18, 1894 (No. 571). *Missouri* or *Flowering Currant.*

Ribes aureum chrysococcus, Rydberg, Fl. Neb. xxi, 71 (1895).

Mr. Rydberg founds this variety upon the character of the fruit. It is what has been called the yellow fruited form of the preceding. Noted by Mr. Rydberg, it may again be stated that there is no discernible difference between the flowers and leaves of the two forms, but the fruits are strikingly different, and apparently the difference is constant. The variety is very common in the Platte canons at Fairbanks and produces large, finely flavored berries in abundance, which are gathered in large quantities for domestic use by the residents of the place.

Fairbanks, July 18, 1894 (No. 571a). *Yellow Currant.*

Ribes cereum, Dougl. Trans. Hort. Soc. Lond. vii, 512 (1830).

Exceedingly common on dry, rocky hillsides throughout the state.

Laramie Hills, May 16, 1894 (No. 30); May, 1895 (No. 1231).

Ribes divaricatum, Dougl. Trans. Hort. Soc. Lond. vii, 515 (1830).

Presumably rare; Union Pass, Aug. 10, 1894 (No. 862).

Ribes divaricatum irrigum, Gray, Am. Naturalist, x, 273.

Growing in great profusion in the copses bordering stream banks and producing an abundance of very acid but pleasantly flavored fruit.

Pole Creek, June 2, 1894 (No. 96). Observed in many other parts of the state, but the borders of Wallace Creek may be spoken of as one immense *Gooseberry* patch.

Ribes floridum, L'Her. Stirp. i, 4 (Ehrb. Beitr. vi. 119).
Observed but once; Whalen Canon, July 18, 1894 (No. 541).

Ribes lacustre, Poir. Encycl. Suppl. ii, 856 (1811).
Very common in copses on canon streams.
Teton Mountains, August 21, 1894 (No. 938); Centennial Valley, fls. June 9, 1895 (No. 1556); fr. August 17, 1895 (No. 1689).

Ribes lacustre parvulum, Gray, Bot. Cal. i, 206 (1876).
Frequent in wet thickets at subalpine stations.
Union Peak, August 13, 1894 (No. 1022); LaPlata Mines, August 23, 1895 (No. 1801).

Ribes lacustre molle, (?) Gray, Bot. Cal. i, 206 (1876).
My specimens are clearly a variety of *R. lacustre*, but as I have no description of this variety I put it here tentatively. It has larger fruits and is more softly pubescent than the species.
Union Pass, August 13, 1894 (No. 1073).

Ribes leptanthum, Gray, Pl. Fendl. 53 (1849).
This horridly prickly form must be very rare in the state for, though it is naturally conspicuous, I have observed it but once.
Garfield Peak, July 29, 1894 (No. 695).

Ribes oxyacanthoides, L. Sp. Pl. 201 (1753).
Collected only in one locality, and probably confined to the eastern part of the state.
Pole Creek, June 2, 1894 (No. 129).

Ribes oxyacanthoides saxosum, (Hook.) Coville, Contrib. Natl. Herb. iv, 100 (1893).
This I at first thought the same as the preceding, but on comparing my specimens with description, l. c., I find that they must be referred to this variety.
Bacon Creek, August 25, 1894 (No. 1044).

Ribes sanguineum variegatum, Wats. Bot. King Surv. 100 (1871).
Union Pass, August 10, 1894 (No. 860).

CRASSULACEÆ.

Sedum Douglasii, Hook. Fl. Bor. Am. i, 228.
  Rare; collected at Wheatland, June 11, 1891, B. C. Buffum.

Sedum rodanthum, Gray, Am. Journ Sci. ii, 33, 405.
  In wet bogs in the mountains; subalpine.
  Union Pass, August 13, 1894 (No. 929); La Plata Mines, August 23, 1895 (No. 1703).

Sedum stenopetalum, Pursh Fl. i, 324 (1814).
  Exceedingly abundant on rocky slopes in the foothills.
  Laramie Hills, July 7, 1894 (No. 351); Table Mountain, June 29, 1895 (No. 1467).

HALORAGEÆ.

Hippuris vulgaris, L. Sp. Pl. 4 (1753).
  Not at all common; in a muddy bog in Lander, August 3, 1894 (No. 694).

Callitriche palustris, L. Sp. Pl. 969 (1753).
  In shallow ponds and pools; rather frequent.
  Near Green River, August 25, 1894 (No. 1102); Centennial Valley, August 25, 1895 (No. 1860).

Myriophyllum verticillatum, L. Sp. Pl. 992 (1753).
  Green River, August 26, 1894 (No. 1038).

ONAGRACEÆ.

Epilobium adenocaulon, Hausskn. Oest. Bot. Zeitschr. xxix (1879).
  Common about springs and wet places.
  Laramie, August, 1891, B. C. Buffum; Whalen Canon, July 19, 1894 (No. 545); Lander Creek, August 31, 1894 (No. 1107).

Epilobium brevistylum, Barbey, Brewer & Wats. Bot. Cal. i, 220 (1876).
  This specimen I had called *E. affine*, but a closer study of text and plate in Rev. Epilobium by Trelease leads me to believe that these specimens belong to the above.
  Wind River Mountains, August 11, 1894 (No. 852).

Epilobium delicatum, Trelease, Mo. Bot. Garden Rep. ii, (1891).
  Of the two specimens before me, the one seems nearly typical, the other is rather large flowered but too immature for close comparison.

Cummins, July 30, 1895 (No. 1466); Bald Mountain, August 15, 1892, B. C. Buffum.

Epilobium Hornemanni, Reichenb. Icon. Crit. ii, 73 (1824).

Again I have two specimens, one typical except that it is very dwarf; the other with leaves sessile instead of short petiolate. However, I am unable to refer them to any other species.

Bald Mountain, August 15, 1892; LaPlata Mines, August 23, 1895 (No. 1797).

Epilobium latifolium, L. Sp. Pl. i, 347 (1753). *Chamænerion latifolium* (L.) Sweet.

A rare and handsome species. The specimens before me may possibly be var. *grandiflorum*, Britton.

On a steep clay bank of the Gros Ventre River, August 16, 1894 (No. 1081).

Epilobium paniculatum, Nutt. T. & G. Fl. i, 490 (1840).

Common on sandy soil among small undershrubs on stream banks.

Wallace Creek, July 28, 1894 (No. 643); Cummins, July 28, 1895 (No. 1456).

Epilobium spicatum, Lam. Fl. Fr. iii, 482 (1778). *Chamænerion angustifolium,* (L.) Scop.

Common in the mountains, especially in districts recently burned over.

Garfield Peak, July 29, 1894 (No. 691). *Fire Weed.*

Epilobium suffruticosum, Nutt.

Large typical specimens were obtained on a dry, rocky creek bed. Warm Spring Creek, August 10, 1894 (No. 798).

Clarkia Pulchella, Pursh, Fl. i, 260, t. 11 (1814).

Probably rare; specimens received from Snake River, May 29, 1892, collected by Fred McCullough.

Gayophytum racemosum, T. & G. Fl. i, 514 (1840).

This and the following species grow in the greatest profusion in the hills, in similar habitats, viz., dry banks, hillsides and ravines.

Sybille Creek, July 8, 1894 (No. 310); Table Mountain, July 1, 1895 (No. 1371); observed in many other localities.

Gayophytum ramosissimum, T. & G. Fl. i, 513 (1840).

Saratoga, June 23, 1893; Sybille Creek, July 8, 1894 (No. 308).

Œnothera albicaulis, Nutt. Fraser's Cat. 1813. *Anogra pallida*, Britton.

Common on sandy ridges along the Laramie River.

Fisher's Ranch, June 19, 1894 (No. 264); Whalen Canon, July 17, 1894 (No. 511).

Œnothera biennis, L. Sp. Pl. 346 (1753). *Onagra biennis* (L.) Scop.

The specimens before me show considerable variation, but none of them seem to fall into any of the existing varieties.

Whalen Canon, July 18, 1894 (No. 519); Laramie Peak, August 8, 1895 (No. 1646).

Œnothera brachycarpa, Gray, Pl. Wright, i, 70.

On gravelly hillsides in the mountains; not frequent.

Centennial Valley, June 9, 1895 (No. 1274); Gros Ventre River, August 16, 1894 (No. 926).

Œnothera breviflora, T. & G. Fl. i, 506 (1840).

This was found in considerable abundance in one spot only, the naked soil of an old buffalo wallow.

Bacon Creek, August 25, 1894 (No. 1043).

Œnothera cæspitosa, Nutt. Gen. i, 246 (1818).

Abundant on the naked red clay hills near Laramie; May 23, 1894 (No. 58); May 30, 1895 (No. 1221).

Œnothera coronopifolia, T. & G. Fl. N. A. i, 495 (1840). *Anogra coronopifolia*, (T. & G.) Britton.

Common on the Laramie Plains.

University Campus, June 1894 (No. 222); June 1895 (No. 1426).

Œnothera Hartwegi, Benth. Pl. Hartw. v, (1843).

On dry, gravelly clay hills; not common.

Blue Grass Creek, July 9, 1894 (No. 375).

Œnothera Hartwegi lavandulæfolia, Watson, Proc. Am. Acad. viii, 590 (1873).

These specimens are from type locality, viz., "Plains of the Platte." Seemingly quite rare.

Whalen Canon, July 18, 1894 (No. 526).

Œnothera pinnatifida, Nutt. Gen. i, 245 (1818). *Anogra albicaulis*, (Pursh) Britton.

Frequent on sandy river bottoms.

Wheatland, July 9, 1894 (No. 380); North Fork Wind River, August 9, 1894 (No. 779).

Œnothera serrulata, Nutt. Gen. i. 246 (1818). *Meriolix serrulata,* (Nutt.) Walp.
　　Infrequent; Orin Junction, August 14, 1894, B. C. Buffum; Whalen Canon, July 18, 1894 (No. 524).
Œnothera trichocalyx, Nutt. T. & G. Fl. i, 494 (1840).
　　Particularly rare in the state; only a few specimens secured.
　　Bull Lake Creek, August 8, 1894 (No. 730).
Œnothera triloba, Nutt. Journ. Acad. Phil. ii, 118 (1821). *Lavauxia triloba,* (Nutt.) Spach.
Gaura coccinea, Nutt. Fraser's Cat. (1813).
　　A common weed everywhere.
　　Uva, July 10, 1894 (No. 367); Laramie, July 24, 1895 (No. 1430).
Gaura parviflora, Dougl. Hook. Fl. Bor. Am. i, 208 (1833).
　　Belonging to the lower altitudes of the state.
　　Two Bar Ranch, Blue Grass Creek, July 9, 1894 (No. 376).

LOASACEÆ.

Mentzelia albicaulis, Dougl. Hook. Fl. Bor. Am. i, 222 (1833).
　　In sandy thickets along streams or on hillsides.
　　Grant, on Sybille Creek, July 8, 1894 (No. 336); Cummins, July 28, 1895 (No. 1471).
Mentzelia chrysantha, Engelm.
　　This is probably rare, as only a few specimens were observed. No citations are at hand; reported from Canon City, Colo., and southern Utah in Coulter's Manual.
　　Big Wind River, August 5, 1894 (No. 705).
Mentzelia dispersa, Watson, Proc. Am. Acad. xi, 115 and 137.
　　Common in dry ravines in the hills.
　　Table Mountain, June 30, 1895 (No. 1375); Cummins, July 28, 1895 (No. 1455).
Mentzelia lævicaulis, T. & G. Fl. N. A. i, 535 (1840).
　　In disintegrated eruptive rock, Garfield Peak, July 29, 1894 (No. 678).
Mentzelia Nelsonii, Greene, Erythea, iii, 70 (1895).
　　The following is the original description: "Annual, 2 or 3 feet high, freely and widely branching, the stoutish branches with a sparingly hispidulous whitish bark; lower leaves unknown, those of the branches from distinctly hastate-ovate to almost deltoid-ovate,

1 or 2 inches long, coarsely toothed or indistinctly lobed, both faces green and rather sparsely appressed-hispidulous, the hairs of the upper surface stouter and more enlarged at base; flowers many, small, orange-colored, sessile, or nearly so, in the forks and axils; ovary subcylindric, less than one-half inch long at flowering time and after; calyx-lobes slenderly subulate at flowering, almost as long as the ovary; petals 5 only, about 4 lines long; stamens few; filaments filiform; anthers suborbicular; capsule and seeds unknown. A very well marked species, certainly allied to the Mexican *M. aspera*, but much larger and more diffusely branching, the leaves relatively broader."

It is probably quite local; collected in a canon leading to the Platte River, July 13, 1894 (No. 439).

Mentzelia nuda, T. & G. Fl. N. A. i, 535 (1840).

Frequent, and always in abundance; sandy plains and hillsides; somewhat variable as to habit.

Big Sandy, July 18, 1892; Grant, July 8, 1894 (No. 338); Cummins, July 29, 1895 (No. 1470).

Mentzelia ornata, T. & G. Fl. N. A. i, 534 (1840).

Also common on sandy foothills near the Platte and its tributaries.

Fairbanks, July 14, 1894 (No. 486); Big Horn Mountains, August 5, 1892.

Mentzelia pumila, T. & G. l. c.

On a stony, gravelly hillside, Cummins, July 29, 1895 (No. 1436).

CACTACEÆ.

It is not at all probable that the following numbers represent at all adequately our *Cactaceæ*, but the difficulty of preparing good specimens has caused them to be neglected.

Cactus viviparus, Nutt. in Fraser's Cat. (1813).

This is exceedingly rare; only two plants thus far observed.

Ione Ranch, on Laramie River, August 10, 1895 (No. 1865).

Echinocactus Simpsoni, Engelm. Trans. St. Louis Acad. ii, 197 (1863).

Frequent on the plains and in the valleys of the Laramie range. Laramie, June 15, 1894 (No. 75).

Cereus viridiflorus, Engelm. Pl. Fendl. 50 (1849).
    Rare, seemingly confined to the east slopes of the Laramie range. Pole Creek, June 2, 1894 (No. 113).

Opuntia fragilis (?) Haw. Suppl. Pl. Succ. 82 (1819).
    A few specimens from Fairbanks, on the Platte, are thought to belong here. July 14, 1894 (No. 465).

Opuntia polyacantha platycarpa, (Engelm.) Coulter, Contrib. Natl. Herb. (III) vii, 436 (1896).
    Probably several varieties of this species are found on our plains, but no specimens are at hand except of this. On the dry, gravelly plains the species, in some form, is immensely abundant, forming in places almost continuous beds.
    Table Mountain, June 2, 1894 (No. 115).

UMBELLIFERÆ.

Sanicula Marylandica, L. Sp. Pl. 235 (1753).
    Rare in the parts of the state collected, but probably frequent in the northeast.
    Laramie Peak, August 8, 1895 (No. 1607).

Musenium tenuifolium, Nutt. T. & G. Fl. N. A. i, 642 (1840). *Adorium tenuifolium,* (Nutt.) Kuntze.
    Usually reported as rare, but it is found in the greatest profusion in the Laramie range, on rocky ridges. June 7, 1894 (No. 176); Platte Hills, July 11, 1894 (No. 391).

Musenium trachyspermum, Nutt. l. c.
    Common on the Laramie plains, appearing very early.
    Laramie, May 7, 1894 (No. 10); observed outside of Laramie but not collected.

Bupleurum Americanum, C. & R.
    Seemingly quite rare; a fine species.
    Union Pass, August 14, 1894 (No. 893).

Bupleurum ranunculoides, L. Sp. Pl. 237 (1753).
    Judging by place of collection, this may be denominated alpine.
    Teton Peaks, August 21, 1894 (No. 972).

Harbouria trachypleura, C. & R.
    Frequent in fertile mountain valleys at 7,000-8,000 ft.
    Table Mountain, June 2, 1894 (No. 160); Saw Mill Creek, May 25, 1895 (No. 1238).

Cicuta maculata, L. Sp. Pl. 256 (1753).
>In wet places along streams, particularly in the lower altitudes.
>>Lusk, July 21, 1894 (No. 573); Meadow Creek, August 9, 1894 (No. 790); Laramie River, near Ione Ranch, August 10, 1895 (No. 1557).

Carum Gairdneri, B. & H. Gen. Pl. i, 891.
>In fertile valleys in the mountains, particularly such as are partially covered with undershrubs. The "Yamp" of the Indians.
>>Garfield Peak, July 29, 1894 (No. 660); Gros Ventre River, August 18, 1894 (No. 1096).

Zizia cordata, DC. Prodr. iv, 100 (1830).
>Along streams and on moist hillsides even to their summits.
>>Horse Creek, June 9, 1894 (No. 204); Pole Creek, June 29, 1895 (No. 1327).

Sium cicutæfolium, Gmelin. Syst. ii, 482 (1791).
>In the margins of fresh water lakes and ponds.
>>Bull Lake, August 8, 1894 (No. 731); Laramie River, August 10, 1895 (No. 1665).

Osmorrhiza nuda, Torr. Pac. Rep. iv, 93 (1857).
>In copses along most of our streams.
>>Garfield Peak, July 29, 1894 (No. 650); Centennial Valley, August 19, 1895 (No. 1722).

Cymopterus montanus, T. & G. Fl. N. A. i, 624 (1840).
>One of the earliest plants on the plains and hills.
>>Laramie, May 7, 1894 (No. 9); specimens from previous years also in the herbarium.

Ligusticum apiifolium. Benth. & Hook. Gen. Pl. i, 912.
>A species with handsome foliage, quite rare.
>>Union Pass, August 14, 1894 (No. 832).

Ligusticum filicinum, Watson, Proc. Am. Acad. xi, 140.
>These specimens were not secured until after most of the fruit had fallen off, but the remaining fruits and foliage made satisfactory determination possible.
>>Gros Ventre river, August 18, 1894 (No. 1095).

Ligusticum scopulorum, Gray, Proc. Am. Acad. vii, 347 (1868).
>Fine specimens were collected at an unusual altitude, nearly 11,000 ft.
>>La Plata Mines, August 23, 1895 (No. 1784).

Ligusticum, sp.
>The plants represented by numbers 1610 and 1655 are too immature to render determination certain, but Dr. Rose, to whom they were submitted, thinks it probable they are distinct species.

Oreoxis humilis, Raf.
>Rare; Cummins, July 30, 1895 (No. 1431).

Selinum Grayi, C. & R.
>In wet places at high elevations.
>La Plata Mines, August 21, 1895 (No. 1776).

Angelica pinnata, Watson, King's Rep. v, 126 (1871).
>Infrequent; along streams at 7,000-8,000 ft.
>Upper Wind River, August 10, 1894 (No. 755).

Peucedanum graveolens, (?) Watson, King's Rep. v, 128 (1871). *P. Kingii,* Wats.
>The material is scanty and over-ripe, but there is little doubt as to the correctness of the determination.
>Garfield Peak, July 29, 1894 (No. 649).

Peucedanum nudicaule, Nutt. T. & G. Fl. N. A. i, 627 (1840).
>Everywhere in the foothills, the naked scapes shooting up almost before the snow is off the ground.
>Laramie, May 4, 1894 (No. 6); also on Horse Creek, June 6, 1893.

Peucedanum simplex, Nutt. Wats. King Rep. v, 129 (1871).
>Only a few specimens secured.
>Union Pass, August 11, 1894 (No. 822).

Pastinaca sativa, L. Sp. Pl. 262 (1753).
>Introduced at Cheyenne, where it was collected by B. C. Buffum, August 11, 1891. *Wild Parsnip.*

Heracleum lanatum, Michx. Fl. Bor. Am. i, 166 (1803).
>On all streams, growing in the greatest profusion in the thickets at the water's edge.
>Horse Creek, July 11, 1891; collected also high up on the Tetons, August 21, 1894 (No. 1055).

ARALIACEÆ.

Aralia hispida, Vent. Hort. Cels. t. 41.
>The herbarium contains a single specimen collected by B. C. Buffum in 1892, no locality given. It most probably is from the north-eastern part of the state.

CORNACEÆ.

Cornus stolonifera, Michx. Fl. i, 92 (1803).
   An exceedingly common shrub in thickets on most of our streams.
   Wallace Creek, July 29, 1894 (No. 663); Little Sandy, August 31, 1894 (No. 1125); Table Mountain, July 1, 1895 (No. 1407).

CAPRIFOLIACEÆ.

Sambucus melanocarpa, Gray, Proc. Am. Acad. xix, 76 (1883).
   Frequent in rocky canons throughout the state.
   Telephone Canon, June 15, 1894, (No. 253); Union Peak, August 13, 1894 (No. 1026); Centennial Valley, August 18, 1895 (No. 1690).

Viburnum pauciflorum, Pylaie, T. & G. Fl. N. A. ii, 17 (1841).
   Reported abundant near Sundance, specimens communicated by Mr. H. J. Chassel, September 1, 1895.

Symphoricarpos occidentalis, Hook. Fl. Bor. Am. i, 285 (1834).
   On sandy creek banks as an undershrub in the thickets.
   Blue Grass Creek, July 8, 1894 (No. 324); Laramie Peak, August 8, 1895 (No. 1565).

Symphoricarpos oreophilus, Gray; Journ. Linn. Soc. & Bot. Calif.
   In the hills and mountains only at considerable elevations.
   Casper Mountain, July 26, 1894 (No. 608); Cummins, July 29, 1895 (No. 1509).

Symphoricarpos racemosus pauciflorus, Robbins, Gray, Man. Ed. 5, 203 (1867). *S. pauciflorus,* (Robbins) Britton.
   A few specimens secured on the eastern slope of the Tetons, August 21, 1894 (No. 958).

Lonicera involucrata, Banks, Richards. Bot. App. Ed. 2, 6 (1823).
   Very common on little mountain streams; frequently called *Grouse Berries.*
   Upper Wind River, August 10, 1894 (No. 758); Cummins, July 30, 1895 (No. 1482).

Lonicera Utahensis, Wats. Bot. King Surv. 133 (1871).
   This I think to be very rare in the state.
   Teton Mountains, August 21, 1884 (No. 934).

RUBIACEÆ.

Galium boreale, L. Sp. Pl. 108 (1753).
  On every fertile mountain hillside and every valley in the greatest profusion.
    Sybille Creek, July 8, 1894 (No. 343); Table Mountain, June 30, 1895 (No. 1384).

Galium trifidum, L. Sp. Pl. 105 (1753).
  Common in wet places, as on the occasionally flooded banks of slow flowing streams.
    Silver Creek, August 24, 1894 (No. 1115); Centennial Valley, August 19, 1895 (No. 1763).

Galium triflorum, Michx. Fl. i, 80 (1803).
  Not common, collected on a wet, shaded hillside.
    Centennial Valley, August 17, 1895 (No. 1693).

VALERIANACEÆ.

Valeriana edulis, Nutt.; Torr & Gray, Fl. ii, 48 (1841).
  Very plentiful in the wet meadows bordering the Laramie River.
    Fisher Ranch, June 19, 1894 (No. 262).

Valeriana Sitchensis, Bong. Veg. Sitch. 145.
  Fine specimens were secured at Clark's, but not observed elsewhere.
    Upper Wind River, August 10, 1894 (No. 793).

Valeriana sylvatica, Banks; Richards Bot. App. 730 (1823).
  On wooded hillsides and in wet valleys in the Laramie Mountains. This species is very abundant.
    Telephone Canon, June 15, 1894 (No. 228). Observed in a large number of other places.

COMPOSITÆ.

Brickellia grandiflora, Nutt. Trans. Am. Phil. Soc. vii, 287 (1841).
  Coleosanthus grandiflorus, (Hook.) Kuntze.
  Frequent on hillsides near the Platte and its tributaries.
    Fairbanks, July 10, 1894 (No. 423); Cummins, July 30, 1895 (No. 1687).

Kuhnia eupatorioides, L. Sp. Pl. Ed. 2. 1662 (1763).
  Infrequent; Laramie, September 1893.

Kuhnia eupatorioides corymbulosa, T. & G. Fl. N. A. ii, 78 (1841).
K. glutinosa, Ell.
> Probably confined to the eastern part of the state.
> Cliffs, near Cold Spring, July 14, 1894 (No. 457).

Liatris punctata, Hook. Fl. Bor. Am. i: 306, t. 55 (1833). *Lacinaria punctata*, (Hook.) Kuntze.
> Abundant in the northern part of Albany and Laramie counties.
> Wheatland, June 30, 1892, B. C. Buffum; Laramie Peak, August 7, 1895 (No. 1564).

Liatris scariosa, Willd. Sp. Pl. iii, 1635 (1804). *Lacinaria scariosa*, (L.) Hill.
> The range of this is about the same as the last but it prefers the rich loam of the valleys.
> Inyan Kara Divide, August 30, 1892, B. C. Buffum; Laramie Peak, August 8, 1895 (No. 1651).

Liatris squarrosa intermedia, D C. Prodr. v, 129 (1836). *Lacinaria squarrosa intermedia*, (Lindl.) Porter.
> Apparently quite local, in the northern part of Laramie county.
> Rawhide Creek, September 4, 1892, B. C. Buffum; Mexican Mines, July 20, 1894 (No. 588).

Gutierrezia Euthamiæ, T. & G. Fl. N. A. ii, 193 (1841). *G. Sarothrae*, (Pursh) Britton & Rusby.
> The most prevalent of our small undershrubs, particularly on the plains.
> University Campus, September 12, 1894 (No. 1133); frequent also on the plains of the Platte.

Grindelia squarrosa, Dunal in D C. Prodr. v, 315 (1836).
> Abundant in all parts of the state thus far traversed.
> Meadow Creek, August 9, 1894 (No. 777); Laramie, September 16, (No. 1148).

Chrysopsis villosa, Nutt. Gen. ii, 151 (1818).
> In this polymorphous species, with so many intermediate forms, it becomes difficult to say which should receive varietal names. Some of the specimens before me, however, are typical.
> Platte River, July 14, 1894 (No. 481); Hartville, July 18, 1894 (No. 585), a very villous form.

Chrysopsis villosa canescens, Gray, Syn. Fl. 123 (1886).
　The commoner form in the north-western part of the state.
　Gros Ventre River, August 16, 1894 (No. 1084); also observed in the Teton Mountains.

Chrysopsis villosa hispida, Gray, Proc. Acad. Phila. 1863, 65.
　This is the form prevalent on the plains about Laramie.
　State Fish Hatchery grounds, July 1891, B. C. Buffum.

Chrysopsis villosa viscida, Gray, Syn. Fl. 123 (1886).
　A characteristic mountain form common in the Medicine Bow range.
　Cummins, July 28, 1895 (No. 1497).

Aplopappus acaulis, Gray, Proc. Am. Acad. vii, 353.
　Frequent and abundant on the Laramie Plains and in the foothills.
　Laramie Hills, June 1893; plains, west of Laramie, June 9, 1895 (No. 1250).

Aplopappus acaulis glabratus, Eaton, Bot. King's Exp. 161.
　Laramie, 1893. The month not noted, but it was probably collected late in the season.

Aplopappus armerioides, Gray, Syn. Fl. i, 132 (1886). *Stenotus armerioides,* Nutt.
　Found only on the "red hills" near Laramie.
　June 15, 1894 (No. 227).

Aplopappus Fremonti, Gray, near var. **Wardi,** Gray, Syn. Fl. i. 128.
　This fine form I at first thought must be a *Bigelovia* as it was rayless, but Mr. M. L. Fernald, who kindly made comparison for me with the specimens in the Harvard herbarium, finds that my specimens correspond closely with Ward's the main difference being that mine have longer pappus.
　Plains, ten miles north of Laramie, August 1, 1895 (No. 1553).

Aplopappus lanceolatus, T. & G. Fl. ii, 241.
　Frequent on grassy slopes and in the valleys of the foothills.
　Laramie, August 1894; Poison Spider Creek, July 27, 1894 (No. 624).

Aplopappus Lyalli, Gray, Proc. Acad. Phila. 1863, 64.
　Quite typical specimens of this fine alpine plant were secured.
　Union Pass, August 13, 1894 (No. 1012).

Aplopappus Nuttallii, T. & G. Fl. N. A. ii, 242 (1842). *Eriocarpum Grindelioides,* Nutt.
> Collected on wet alkali flats.
> Laramie, October 1893.

Aplopappus Parryi, Gray, Am. Journ Sci. Ser. 2, xxxiii, 10.
> In partially shaded woods among the fallen trees.
> Centennial Hills, August 17, 1895 (No. 1695).

Aplopappus pygmæus, Gray, Am. Journ. Sci. Ser. 2, xxxiii, 239.
> Sparingly found on the bleak, rocky summits of the Medicine Bow Mountains, dwarf and somewhat cæspitose.
> La Plata Mines, August 23, 1895 (No. 1875).

Aplopappus spinulosus, D C. Prodr. v, 347 (1836). *Eriocarpum spinulosum,* (Pursh) Greene.
> Very abundant, in some localities becoming a weed.
> Sheridan Experiment Farm, September 1895, J. F. Lewis; Cold Springs, July 14, 1894 (No. 456).

Aplopappus unifloras, T. & G. Fl. ii, 241.
> So far found only in the north-western part of the state.
> Bacon Creek, August 15, 1894 (No. 911); Green River, August 26, 1894 (No. 1035).

*****Bigelovia collinus,** (Greene). *Chrysothamnus collinus,* Greene.
> For this specimen I am indebted to Prof. Greene. It is quite distinct from all the other forms I have secured.
> Rock Springs, August 9, 1895, Prof. E. L. Greene.

Bigelovia Douglasii, Gray, Proc. Am. Acad. viii, 645 (1873).
> The determination was made by Dr. Rose, but it should be added that the specimens are somewhat immature, and, as stated by him, not in condition to determine with certainty.
> Laramie, 1893.

Bigelovia Douglasii lanceolata, Gray, Syn. Fl. 140 (1886).
> I judge the specimens before me to be nearly typical; they come from within the type locality as well.
> Union Pass, August 14, 1894 (No. 889); also from Bacon Creek.

---

*The nomenclature of this genus is in such a state of confusion, that for the present I adopt that which allows of the quickest and easiest disposition of my material. My library facilities are too meager for me to presume to pass upon the relative merits of Dr. Gray's Bigelovia, Dr. Britton's Chondrophora, and Prof. Greene's Chrysothamnus.

Bigelovia Douglasii pumilla, Gray, Syn. Fl. 140 (1886).
   Of frequent occurrence and apparently throughout the state.
   In 1894 successively at Garfield Peak, Bacon Creek, Boulder Creek and at Laramie, (Nos. 616, 902, 1121, 1197).

Bigelovia Douglasii Stenophylla, Gray, Proc. Am. Acad. viii, 644 (1873).
   Certainly very rare; noted but once.
   Centennial Valley, August 26, 1895 (No. 1847).

Bigelovia Douglasii tortifolia, Gray, l. c.
   These varieties run so closely together that it is difficult to speak with certainty regarding them.
   Poison Spider Creek, July 26, 1894 (No. 617).

Bigelovia graveolens, Gray, l. c.
   Good specimens of this were obtained on the Platte. Common in the canons and foothills near the river.
   Platte River July 14, 1894 (No. 503).

Bigelovia graveolens albicaulis, Gray, l. c.   *Chrysothamnus frigidus,* Greene, Erythea, iii, 112 (1895).
   This is by far the most abundant form on the Laramie Plains, where in places it forms an almost uninterrupted growth for miles at a stretch.
   Its characteristics are so well marked that Prof. Greene is well justified in raising it to specific rank.
   Laramie, August 29, 1891, B. C. Buffum; E. L. Greene, August 1895; Bacon Creek, August 15, 1894 (No. 910).

Bigelovia graveolens glabrata, Gray, l. c.
   This variety was observed only on the Pacific slope.
   Bacon Creek, August 23, 1894 (No 966); Boulder Creek, August 26, (No. 1120).

Bigelovia Howardii, Gray, Proc. Am. Acad. viii, 644 (1873).
   Prof. Greene * notes this form as peculiar to mountain parks of Colorado. This was secured in a similar location in Wyoming.
   Centennial Valley, August 26, 1895 (No. 1846).

Bigelovia linifolia, (Greene).   *Chrysothamnus linifolius,* Greene.
   For this specimen also I am indebted to Prof. Greene; however, on comparing with our material, I find one listed as *B. Douglasii*

---
* Erythea, iii, 114.

lanceolata that perfectly accords with it. This I cut out and place here

Rock Springs, August 9, 1895, E. L. Greene; Poison Spider Creek, July 26, 1894 (No. 618).

Solidago Canadensis, L. Sp. Pl. 878 (1753).

Probably confined to the lower altitudes of the eastern part of the state.

C. Y. Ranch, Big Muddy, July 23, 1894 (No. 597).

Solidago elongata, Nutt. Trans. Am. Phil. Soc. n. ser. vii, 328 (1841).

Frequent on the lower courses of mountain streams.

Big Muddy Creek, July 24, 1894 (No. 641); Cummins, July 27, (No. 1479). Also observed on Meadow Creek, 1894.

Solidago humilis nana, Gray, Proc. Am. Acad. viii, 389.

Infrequent, 9,000 ft. and upward.

Centennial Valley, August 18, 1895 (No. 1680).

Solidago Missouriensis, Nutt. Journ. Acad. Phila. vii, 32 (1834).

The species is much rarer with us than the following varieties.

Laramie Peak, August 8, 1895 (No. 1629).

Solidago Missouriensis extraria, Gray, Proc. Am. Acad. xvii, 195.

This seems to be the form on the western slope of the Rockies.

Bacon Creek, August 15, 1894 (No. 912).

Solidago Missouriensis montana, Gray, l. c.

Very common in dry, clayey ravines and on the adjoining hillsides near the Platte and its tributaries.

Uva, July 10, 1894 (No. 382); Cottonwood Canon, August 5, 1895 (No. 1571).

Solidago multiradiata, Ait. Hort. Kew. iii, 218 (1789).

Frequent in the mountains.

Warm Spring Creek, August 10, 1894 (No. 800); LaPlata Mines, August 22, 1895 (No. 1771).

Solidago multiradiata scopulorum, Gray, Proc. Am. Acad. xvii, 191 (1882).

Only a few specimens were secured, at high elevations, probably 10,000 ft.

Teton Mountains, August 22, 1894 (No. 955).

Solidago nana, Nutt. Trans. Am. Phil. Soc. vii, 327.

Not frequent. Upper Wind River, August 10, 1894 (No. 765).

Solidago rigida, L. Sp. Pl. 880 (1753).

Not observed by the writer, but good specimens from the northeastern part of the state.

Suggs Road, August 15, 1892, B. C. Buffum; Sheridan Experiment Farm, September 1895, J. F. Lewis.

Townsendia grandiflora, Nutt. Trans. Am. Phil. Soc. n. ser. vii, 306 (1841).

Frequent on the sandy plains in the eastern part of the state.

Uva, July 9, 1894 (No. 385); Pole Creek, June 30, 1895 (No. 1366).

Townsendia sericea, Hook. Fl. Bor. Am. 119 (1834). *T. exscapa,* (Richards) Porter.

So far as my observation goes, this is the very earliest flower of southeastern Wyoming. Abundant on the plains and in the foothills.

Collected May 5, 1894 (No. 7); observed on several years as early as the first week in April.

Aster adscendens, Lindl. Hook. Fl. Bor. Am. ii, 8 (1834).

I find this as variable as it is common in the state.

Nearly typical specimens from Bacon Creek, Silver Creek, Sweetwater River and Laramie, late August and September (Nos. 1652, 1110, 1196 and 1149).

A very peculiar form from Myersville, September 5, 1894 (No. 1193). This will probably prove to be a good variety at least. Other specimens from Laramie Peak are not typical, but for the present they are placed here. (Nos. 1561 and 1639).

Aster adscendens frondeus, Gray.

The citation for this name, which was communicated to me by Prof. Greene, I am unable to give. The specimens indicate a good variety at least.

Bacon Creek, August 25, 1894 (No. 1049).

Aster adscendens, Lindl. var. ?

Three quite different forms I have listed as possible varieties— No. 892 from Bacon Creek, Nos. 964 and 1092 from Gros Ventre River.

Aster canescens, Pursh. Fl. Am. Sept. 547 (1814). *Machaeranthera canescens,* Gray.

The specimens seem hardly typical, but neither are they referable to any of the established varieties.

Bacon Creek, August 15, 1894 (No. 904); Gros Ventre River, August 17, 1894 (No. 1089).

Aster canescens latifolius, Gray, Torr. in Emory Rep. 141.

Infrequent; on eastern slope of partially wooded mountain side. Laramie Peak, August 7, 1895 (No. 1636).

Aster canescens viridis, Gray, Syn. Fl. 206 (1886).

A very abundant plant on the Laramie Plains, in places a troublesome weed.

University Campus September 15, 1894 (No. 1150).

Aster commutatus, Gray, Syn. Fl. 185 (1884). *A. incanopilosus,* (Lindl.) Sheldon.

Exceedingly abundant everywhere, some of its forms shading off into *A. multiflorus.*

Big Muddy Creek, July 23, 1894 (No. 598); Laramie, October 10, 1894 (No. 1170).

Aster conspicuus, Lindl. Hook. Fl. Bor. Am. ii, 7 (1840).

Infrequent; on a partially wooded slope in the Gros Ventre Mountains, near the Gros Ventre River, August 22, 1894 (No. 1069).

Aster elegans, T. & G. Fl. N. A. ii, 159.

This beautiful species seems to belong to the northeastern part of the state, in partially open hillsides.

Union Pass, August 11, 1894 (No. 826); Gros Ventre, August 22 (No. 1068).

Aster Engelmanni, Gray, Bot. King Surv. 144 (1871).

In a rather open copse on the banks of a canon stream. Centennial Valley, August 17, 1895 (No. 1691).

Aster foliaceus, Lindl. DC. Prodr. v, 228 (1836).

The specimens are typical except that they are somewhat dwarf. Wheatland, July 14, 1894, B. C. Buffum.

Aster foliaceus Burkei, Gray, Syn. Fl. 193 (1884).

Three good varieties of this at hand, this one from Little Sandy Creek, August 30, 1894 (No. 1132).

Aster foliaceus Eatoni, Gray, Syn. Fl. 194 (1884).

East Fork, August 25, 1894 (No. 1118).

Aster foliaceus frondeus, Gray, Syn. Fl. 193 (1884).
   Centennial Valley, August 25, 1895 (No. 1859).
Aster Fremontii, Gray, T. & G. Fl. N. A. ii, 503.
   No data on this specimen except collected near Laramie, October 3, 1891, by B. C. Buffum.
Aster frondosus, T. & G. Fl. N. A. ii, 165.   *Brachyactis frondosa,* Gray.
   Frequent in low, wet ground, especially in alkali soil, on the hummocks in alkali bogs.
   Typical specimens from Laramie, September 7, 1895 (No. 1867); specimens from Seven Mile Lake, October 15, 1894 (No. 1159), remarkable for their size and the great number of large flowers.
Aster glaucus, T. & G. Fl. N. A. ii, 172.
   In the foothills of all the mountain ranges yet collected.
   Big Wind River, August 10, 1894 (No. 772); Laramie, September 30, 1894 (No. 1151); Laramie Peak, August 7, 1895 (No. 1590).
Aster lævis Geyeri, Gray, Syn. Fl. 183 (1884).
   Collected by B. C. Buffum, Eagle Rock Canon, August 22, 1892.
Aster integrifolius, Nutt. Trans. Am. Phil. Soc. n. ser. vii, 291 (1841).
   Belonging to the western or Pacific slope, observed only on the west side of Union Pass, August 13, 1894 (No. 1032).
Aster Lindleyanus, T. & G. Fl. N. A. ii, 122.
   It was quite a surprise to find this species so far to the south and west.
   Laramie Peak, August 6, 1895 (No. 1592).
Aster multiflorus, Ait. Hort. Kew. iii, 203 (1789).
   Common throughout the state and quite variable as to habit, hirsuteness and size of flowers.
   Big Muddy Creek, July 24, 1894 (No. 642); reported from Sheridan as a weed on the Experiment Farm.
Aster Parryi, Gray, Am. Nat. viii, 212.
   This beautiful large flowered species is found in the greatest abundance in some parts of the Laramie Plains as is its congener *A. xylorrhiza.* Usually in separate districts but occasionally striving for the occupancy of the same hillside. Hybrids, I think, sometimes occur, for specimens are found with the characters of both species well blended. Preferring a red clay soil with a percentage of alkali in it.

Carbon, June 18, 1894 (No. 256); Miss Lily Boyd; Laramie Plains, June 20, 1895 (No. 1315).

Aster ptarmicoides, T. & G. Fl. N. A. ii, 160 (1841).
Rare; typical specimens from Bald Mountain, August 16, 1892, B. C. Buffum.

Aster pulchellus, Eaton, Bot. King Exp. 143.
This elegant alpine form occurs at least in the two principal ranges in the northern part of the state.
Bald Mountain, August 8, 1892, B. C. Buffum; Union Peak, August 13, 1894 (No. 1016).

Aster salicifolius, Ait. Hort. Kew. iii, 203.
From Jackson's Hole, near the base of the Teton Mountains, August 21, 1894 (No. 1065).

Aster tanacetifolius, H B K. Nov. Gen. & Sp. iv, 95 (1820).
Frequent on the sandy plains of the Platte. Uva, July 10, 1894 (No. 443). Observed in many other places.

Aster xylorrhiza, T. & G. Fl. N. A. ii, 158
This grows even more profusely than the above mentioned *A. Parryi*, but prefers soil with less alkali. Sandy clay ridges on the height of land between streams.
Laramie Plains, June 20, 1895 (No. 1315).

Erigeron acris, L. Sp. Pl. 863 (1753).
Infrequent and alpine; Laramie Peak, August 7, 1895 (No. 1621).

Erigeron armerifolius, D C. Prodr. v, 291 (1836).
Quite frequent along stream banks.
Head of Green River, August 14, 1894 (No. 908); Cummins, July 29, 1895 (No. 1480).

Erigeron cæspitosus, Nutt. Trans. Am. Phil. Soc. vii, 307 (1841).
Frequent on dry stony hillsides in the Laramie Mountains.
Blue Grass Creek, July 9, 1894 (No. 334); Cummins, July 30, 1895 (No. 1496).

Erigeron Canadensis, L. Sp, Pl. 863 (1753).
Sybille Creek, July 9, 1894 (No. 298); Whalen Canon, July 18, 1894 (No. 556).

Erigeron canus, Gray, Mem. Am. Acad. iv, 67 (1849).
My specimens of this are a little scanty and over ripe, but I think there is little doubt of the correctness of the determination.
Platte Canon, July 14, 1894 (No. 482).

Erigeron compositus, Pursh, Fl. ii, 535 (1814).

Observed only on a clayey, gravelly ridge at the head of Pole Creek, May 12, 1894 (No. 26); May 18, 1895 (No. 1217).

Erigeron compositus pinnatisectus, Gray, Proc. Am. Acad. xvi, 90 (1880).

In the Medicine Bow Mountains only as yet, 10,000 ft. and upward.

La Plata Mines, August 23, 1895 (No. 1816).

Erigeron trifidus, Hook. Fl. Bor. Am. ii, 17 (1834).

Near the summit of Laramie Peak, August 7, 1895 (No. 1612).

Erigeron corymbosus, Nutt. Trans. Am. Phil. Soc. vii, 308 (1841).

This fine mountain form is quite variable as to the number of heads. In three specimens from different localities I find the stems unicephalous, while only one has the typical corymbose arrangement.

Telephone Canon, June 15, 1894 (No. 234); Platte Hills, July 24, 1894 (No. 633); Union Pass, August 10, 1894 (No. 859); Centennial Hills, June 9, 1895 (No. 1200).

Erigeron Coulteri, Porter, in Fl. of Colo., 61.

Just a few fine specimens secured at Cummins, July 28, 1895 (No. 1524).

Erigeron divergens, T. & G. Fl. ii, 175 (1841).

Widely distributed, but not common.

Fine specimens from Snake River, Jackson's Hole, August 22, 1894 (No. 1053). Larger and more divergently branched plants from subalpine slopes on Laramie Peak, August 6, 1895 (No. 1635).

Erigeron flagellaris, Gray Mem. Am. Acad. iv, 68 (1849).

Immature and mature specimens present a very different appearance on account of the great lengthening of the stems without a corresponding increase in the number of leaves. Table Mountain, June 30, 1895 (No. 1386); Laramie Peak, August 6, 1895 (No. 1600).

Erigeron glabellus, Nutt. Gen. ii, 147 (1818). *E. asperus,* Nutt.

Our commonest Erigeron, everywhere abundant in wet meadows, somewhat variable.

Lander Creek, August 29, 1894 (No. 1117); Cummins, July 30, 1895 (Nos. 1454 and 1536). Specimens from several other localities.

Erigeron subtrinervis, Rydberg.   *E. glabellus mollis*, Gray.
That this is worthy of the specific rank recently accorded it, I think no one will question.
Centennial Valley, August 17, 1895 (No. 1692).

Erigeron grandiflorus, Hook. Fl. ii, 123.
Bald Mountain, August 15, 1892, B. C. Buffum, specimens with rays almost white; La Plata Mines, August 24, 1895 (No. 1805), rays purple.

Erigeron leiomeris, Gray, Syn. Fl. 211 (1884).
Collected on the Grand Teton at about 10,000 ft., at the foot of rocky ledges, August 21, 1894 (No. 1054).

Erigeron macranthus, Nutt. Trans. Am. Phil. Soc. vii, 310 (1841).
In mountain parks and meadows at 8,000 ft. and upward.
Garfield Peak, July 29, 1894 (No. 647); Cummins, July 30, 1895 (No. 1535).

Erigeron pumilus, Nutt. Gen. ii, 147 (1818).
Frequent in sandy, grassy valleys in the Laramie Mountains.
Laramie Hills, June 7, 1894 (No. 169); Table Mountain, June 30, 1895 (No. 1339); observed in many other localities.

Erigeron radicatus, Hook. Fl. ii, 17.
On dry, stony ridges and subalpine table lands.
State Fish Hatchery grounds, Laramie, May 28, 1892, B. C. Buffum; Table Mountain, June 2, 1894 (No. 143).

Erigeron salsuginosus, Gray, Proc. Am. Acad. xvi, 93 (1880).
Wyoming must be the natural home of this splendid species, judging by the luxuriance of its growth. Superb specimens with heads two inches in diameter are of frequent occurrence along our mountain streams.
Union Pass, August 12, 1894 (No. 895); Centennial Hills, August 16, 1895 (No. 1775).

Erigeron strigosus, Muhl. Willd. Sp. Pl. iii, 1956 (1804).   *E. ramosus*, (Walt.) B. S. P.
Infrequent, Union Pass, August 11, 1894 (No. 851).

Erigeron uniflorus, L. Sp. Pl. 864 (1753).
This, with us, alpine form varies greatly as to size and hirsuteness.
Specimens from Teton Mountains, August 22, 1894 (No. 969), are only 1-2 inches high, the involucre hirsute with sparse light

colored hairs. Those from the Medicine Bow Mountains, August 23, 1895 (No. 1772) are 4-6 inches high, stems bearing 4-6 leaves; involucre densely black lanate, rays white, heads 1 inch in diameter. I suggest the varietal name **melanocephalus** for this form.

Erigeron ursinus, Eaton, Bot. King's Exp. 148 (1871).

On rocky hills and ledges, 7,000-11,000 ft., successively throughout the season.

Table Mountain, June 2, 1894 (No. 144); La Plata Mines, August 23, 1895 (No. 1795).

Filago depressa, Gray, Proc. Am. Acad. xix, 3.

Not having seen an authentic *F. depressa*, I give the above determination with some reservation. In a gravelly hollow near the Little Laramie River, Centennial Valley, August 19, 1895 (No. 1751).

Antennaria alpina, Gaertn. Fr. & Sem. ii, 410 (1791).

Frequent, on dry hillsides.

Union Pass, August 11, 1894 (No. 853); Centennial Valley, June 9, 1895 (No. 1265).

Antennaria Carpathica pulcherrima, Hook. Fl. i, 329 (1834).

Probably throughout the state, in the rich soil of thickets.

Pole Creek, June 9, 1894 (No 110); Union Pass, August 12, 1894 (No. 819).

Antennaria dioica, Gaertn. l. c.

Of this variable species we have our full share of forms. Laramie, Inyan Kara Divide, Wind River and Green River are places from which specimens are at hand (Nos. 762 and 885).

Antennaria dioica congesta. D C. Prodr. vi, 269.

This is of frequent occurrence, sometimes with closely depressed stems, at other times, stems several inches in length. Separable from the species by the compactness of the heads.

Laramie, June 28, 1894 (No. 291). A form with strikingly roseate bracts probably belongs here.

Antennaria racemosa, Hook. Fl. Bor. Am. i, 330 (1834).

Rare, and but a few specimens secured.

Union Pass, August 11, 1894 (No. 812).

Anaphalis margaritacea, Benth. & Hook. Gen. Pl. ii, 303 (1873).

Antennaria margaritacea, (L.) Hook.

In deeply shaded copses at the foot of mountains.

Tetons, August 21, 1894 (No. 959); Laramie Peak, August 8, 1895 (No. 1604).

Anaphalis margaritacea subalpina, Gray, Syn. Fl. 233 (1884).
On a rocky creek bed, Centennial Valley, August 16, 1895 (No. 1669).

Gnaphalium palustre. Nutt. Trans. Am. Phil. Soc. n. ser. vii, 404 (1841).
Not common. Wheatland, August 8, 1891, B. C. Buffum; Atlantic City, September 3, 1894 (No. 1186).

Iva axillaris, Pursh, Fl. Am. Sept. 743 (1814).
A troublesome weed in some localities; frequent on the sandy plains of the Platte.
Fairbanks, July 14, 1894 (No. 476). *Poverty Weed.*

Iva xanthifolia, Nutt. Trans. Am. Phil. Soc. (II) vii, 347 (1841).
Infrequent; Willow Creek, July 20, 1894 (No. 569); Cheyenne, August 11, 1891, B. C. Buffum.

Ambrosia psilostachya, D C. Prodr. v, 526 (1836).
An exceedingly annoying weed, its root-stocks making it almost impossible to destroy it by cultivation.
Wheatland, August 11, 1891, B. C. Buffum ; Fairbanks, July 11, 1894 (No. 425); Laramie, September 30, 1894.

Ambrosia trifida, L. Sp. Pl. 987 (1753).
Infrequent in the state as yet.
Ford J. Ranch, on Willow Creek, July 21, 1894 (No. 563).

Franseria discolor, Nutt. Trans. Am. Phil. Soc. (II) vii, 345 (1841).
Gærtneria discolor, (Nutt.) Kuntze.
One more of our weeds; "the more it is dug up the better it thrives."
University campus, July, 1891, B. C. Buffum ; Hartville, July 15, 1894 (No. 550).

Franseria Hookeriana, Nutt. l. c. *Gærtneria acanthicarpa,* (Hook.) Britton.
Frequent on sandy plains.
Inyan Kara Divide, August 30, 1892, B. C. Buffum ; Big Wind River, August 5, 1894 (No. 707); Sweetwater River, September 9, 1894 (No. 1190).

Xanthium Canadense, Mill. Gard. Dict. Ed. 8, No. 2 (1768).
   Not frequent; Cheyenne, August 9, 1891, B. C. Buffum; Platte River, July 14, 1894 (No. 485).

Gymnolomia multiflora, Benth. & Hook. Rothr. Bot. Wheeler Surv. 160 (1876).
   Only from Snake River thus far, August 21, 1894 (No. 1064).

Rudbeckia hirta, L. Sp. Pl. 907 (1753).
   Common throughout the state.
   Specimens from Cheyenne, Laramie, Big Muddy Creek, and Cummins, (Nos. 600, 1144, 1459).

Rudbeckia laciniata, L. Sp. Pl. 906 (1753).
   Common on streams in eastern part of the state.
   Cottonwood Canon, August 5, 1895 (No. 1575).

Lepachys columnaris, T. & G. Fl. N. A. ii, 314 (1842).
   This and the following variety are common on the plains and hills about the Platte
   Uva, July 10, 1894 (No. 386).

Lepachys columnaris pulcherrima, T. & G. l. c.
   This form maintains itself pretty uniformly in given areas, so seems entitled to varietal name.
   Willow Creek, July 20, 1894 (No. 570).

Balsamorrhiza sagittata, Nutt. Trans. Am. Phil. Soc. vii, 349 (1841).
   Occasionally growing in the greatest profusion in dry, stony ravines.
   Laramie Hills, by B. C. Buffum, June 24, 1892; June 9, 1894 (No. 213).

Balsamorrhiza macrophylla, Nutt. Trans. Am. Phil. Soc. vii, 350 (1841).
   Very rare, but one specimen found.
   Union Pass, August 10, 1894 (No. 921).

Wyethia amplexicaulis, Nutt. Trans. Am. Phil. Soc. vii, 349 (1841).
   Infrequent; Laramie Hills, June 21, 1891, B. C. Buffum; Union Pass Hills, August 11, 1894 (No. 816).

Helianthus annuus, L. Sp. Pl. ii, 904 (1753).
   A common weed in some localities.
   Cheyenne, August 11, 1891, B. C. Buffum; Fairbanks, July 12, 1894 (No. 432).

Helianthus giganteus, L. Sp. Pl. 905 (1753).
   Near streams and in wet bottoms; not frequent.
   Muskrat Creek, July 30, 1894 (No. 684); Laramie, October 6, 1894 (No. 1169).

Helianthus Nuttallii, T. & G. Fl. N. A. ii, 324 (1842).
   Secured only west of the Wind River Mountains.
   Gros Ventre River, August 16, 1894 (No. 1083).

Helianthus pumilus, Nutt. Trans. Am. Phil. Soc. vii, 366.
   Frequent on dry slopes and hilltops.
   Platte River Hills, July 14, 1894 (No. 501); also from Casper.

Helianthus rigidus, Desf. Cat. Hort. Paris. Ed. 3, 184 (1813).
   Whalen Canon, July 18, 1894 (No. 528); Laramie Peak, August 6, 1895 (No. 1578).

Helianthella quinquenervis, Gray, Proc. Am. Acad. xix, 10.
   In the mountains, in open places near streams.
   Bald Mountain, by B. C. Buffum; Union Pass, August 10, 1894 (No. 803); Laramie Peak, August 6, 1895 (No. 1654).

Thelesperma gracile, Gray, Kew. Journ. Bot. i, 252 (1849).
   Frequent on the hills near the Platte Riiver.
   Near Fort Laramie by B. C. Buffum; Blue Grass Hills, July 8, 1894 (No. 321).

Bidens frondosa, L. Sp. Pl. 832 (1753).
   Frequent in the eastern part of the state.
   Wheatland, September 1892; Sheridan Experiment Farm, September 1895, J. F. Lewis.

Bidens cernua, L. Sp. Pl. 832 (1753).
   In wet places about Wheatland, September 1892, B. C. Buffum.

Madia glomerata, Hook. Fl. ii, 24.
   A rare plant; in dry rich loam soil.
   Big Wind River, August 2, 1892, B. C. Buffum; Laramie Peak, August 8, 1895 (No. 1593).

Chænactis Douglasii, Hook. & Arn. Bot. Beechy. 354 (1840-41).
   On abrupt stony clay hills.
   Mouth of Bacon Creek, August 15, 1894 (No. 909); Cummins, July 25, 1895 (No. 1439).

Chænactis Douglasii alpina, Gray, Syn. Fl. 341 (1884).
   On subalpine stony hilltops; easily distinguished from the species by its striking rosette of radical leaves and its smaller size.

Garfield Peak, July 29, 1894 (No. 653); Cummins, July 26, 1895 (No. 1438).

Hymenopappus filifolius, Hook. Fl. Bor. Am. i, 317 (1833).

Frequent on the dry slopes of the Platte and Laramie River hills at 5,000-7,000 ft.

Sybille Creek, July 7, 1894 (No. 328); Table Mountain, June 29, 1895 (No. 1369).

Hymenopappus ligulæflorus, n. sp.

Perennial from a multicipital caudex, each division bearing one leafy stem, 5-8 inches high, glabrous but for some floccose wool on the crown of caudex; leaves simply pinnate into about five linear divisions, impressed punctate; heads few, corymbosely cymose, about $\frac{1}{4}$ inch high; involucral bracts oblong, hirsute-villous on the margins, resinous-atomiferous as are also the corollas, the whole strong scented; rays 6-8, $\frac{1}{4}$ inch long; lobes of disk corollas very short and erect; achenes short villous; pappus of thin acuminate paleæ, in this respect allying it more closely to *Hymenothrix*; flowers yellow.

Mr. L. Fernald, assistant in Gray Herbarium, who kindly examined it for me, reports it as a form of *H. filifolius*, but I cannot see why it should be left there.

Collected on the north Laramie Plains, about six miles from Owen, August 5, 1895 (No. 1573).

Bahia chrysanthemoides, Gray, Proc. Am. Acad. xix, 28.

Noted at Laramie Peak only, August 6, 1895 (No. 1634).

Bahia oppositifolia, Nutt. T. & G. Fl. N. A. ii, 376 (1842).

Common on the dry, sandy plains of the Platte; July 1894 (Nos. 332 and 602).

Eriophyllum cæspitosum, (Dougl.) Lindl. Bot. Reg. xiv, 1167 (1828).

Probably confined to the northwestern part of the state.

B. C. Buffum, in 1892 without data; Gros Ventre River, August 18, 1894 (No. 1099).

Eriophyllum cæspitosum leucophyllum, Gray, Proc. Am. Acad. xix, 25.

Very rare, in dry stony ravine.

Warm Spring Creek, Union Pass, August 10, 1894 (No. 801).

Dysodia chrysanthemoides, Lag. Nov. Gen. et Sp. 29 (1816). *D. papposa*, (Vent.) A. S. Hitchc.
Common in the eastern part of the state.
From Wheatland, by B. C. Buffum, August 11, 1891; Platte River, July 14, 1894 (No. 499).

Helenium autumnale, L. Sp. Pl. 886 (1753).
Very frequent on river bottoms.
Popo Agie River, August 1, 1894 (No. 736); Ione Ranch, August 10, 1895 (No. 1664).

Helenium Hoopesii, Gray, Proc. Acad. Phil. 65 (1863).
Infrequent; Union Pass, August 11, 1894 (No. 841).

Gaillardia aristata, Pursh, Fl. ii, 573.
Frequent and quite variable especially as to foliage. Probably throughout the state.
Platte River Hills, July 11, 1894 (No. 417); Pole Creek, June 27, 1895 (No. 1326).

Actinella acaulis, Nutt. Gen. ii, 173.
Very frequent indeed in various forms, the earlier individuals scapeless as well as stemless.
Laramie Hills, June 7, 1894 (No. 177); Table Mountain, June 27, 1895 (No. 1300).

Actinella glabra, (Nutt.) n. sp.
If ever a plant deserved specific rank, this one does. Many as are the forms of *A. acaulis*, by necessity, there is no excuse for making this one of them. It is clearly separated from that by the much longer branches of the caudex, which are closely covered with the persistent bases of dead petioles, all of which are completely enveloped in long, densely matted wool, brownish-red, except at the summit, where it becomes continuous with the white persistent wool of the scape and involucre. Leaves longer, glabrous and strongly impressed-punctate; heads large, 1 inch or even more across. Roots enormous, sometimes several feet in length; the multicipital caudex forming raised rounded tufts 6-12 inches across, which very early in the spring become covered with fine yellow heads. *A. acaulis glabra*, Gray. *A. glabra*, Nutt.
Exceedingly abundant in the Laramie Hills, where it is in blossom from April to June (Nos. 36 and 1233).

Actinella grandiflora, T. & G. Bost. Journ. Nat. Hist. Soc. v, 110.
  Rare; probably strictly alpine.
  On the naked summits of the Medicine Bow Mountains, August 23, 1895 (No. 1822).

Actinella Richardsonii, Nutt. Trans. Am. Phil. Soc. vii, 379.
  Rare; only a few specimens secured.
  Centennial Hills, August 19, 1895 (No. 1688).

Actinella scaposa linearis, Nutt. Trans. Am. Phil. Soc. vii, 378.
  This is a common species in sandy, grassy valleys in the hills.
  Table Mountain, June 2, 1894 (No. 91); June 29, 1895 (No. 1341).

Achillea millefolium, L. Sp. Pl. ii, 899 (1753).
  From the summits of our mountains to our lowest valleys.
  Sybille Creek, July 8, 1894 (No. 409); Wind River Mountains, August 8, 1894.

Anthemis cotula, L. Sp. Pl. 894 (1753).
  Introduced in Laramie and possibly elsewhere but apparently not thriving. It would soon disappear were in not reintroduced.
  Laramie, October 1894 (No. 1160).

Tanacetum capitatum, T. & G. Fl. N. A. ii, 415.
  Infrequent; Laramie, in the "red hills," June 15, 1894 (No. 226).

Artemisia biennis, Willd. Phytogr. 11 (1794).
  Very common in waste ground about the city.
  Laramie, September 18, 1894 (No. 1145).

Artemisia cana, Pursh, Fl. ii, 521.
  This seems to belong to the western slope, where, in the fertile creek valleys, it is the prevailing species.
  Bacon Creek, August 25, 1894 (No. 1042); Lewiston, September 5, 1894 (No. 1183).

Artemisia Canadensis, Michx. Fl. Bor. Am. ii, 129 (1803).
  Of frequent occurrence on the Big Wind River; by B. C. Buffum, August 2, 1892; August 5, 1894 (No. 710); Laramie Peak, August 6, 1895 (No. 1660).

Artemisia dracunculoides, Pursh, Fl. Am. Sept. 742 (1814).
  Infrequent in the parts of the state collected; Eagle Rock Canon, August 22, 1892, B. C. Buffum.

Artemisia filifolia, Torr. Ann. Lyc. N. Y. ii. 211 (1827).

This fine species is also rare unless it be in the northeastern part of the state.

Inyan Kara Divide, August 30, 1892, B. C. Beffum.

Artemisia frigida, Willd. Sp. Pl. iii, 1838 (1804).

Perhaps the commonest of our long list of "sages," appearing everywhere on the dry plains and in the hills.

Union Pass, August 10, 1894 (No. 861); University campus, September 16, 1894 (No. 1135); Sheridan Experiment Farm, September 1895.

Artemisia Ludoviciana, Nutt. Gen. ii, 143 (1818).

Quite variable, differing especially as to foliage and compactness of panicle. On creek banks throughout the state.

Laramie, October 6, 1894 (No. 1171); Laramie Peak, August 5, 1895 (No. 1643).

Artemisia Ludoviciana integrifolia, n. var.

Leaves all entire, large, (1-3 inches long), narrowly lanceolate, margins revolute; panicle strict, heads fewer and larger than in the species.

Willow Creek, July 20, 1894 (No. 568).

Artemisia Mexicana, Willd. Spreng. Syst. iii, 490.

Infrequent; Sweetwater Stage Station, September 9, 1894 (No. 1181).

Artemisia Norvegica, Fries, in Liljeb. Fl. (1815).

A handsome plant, infrequent, alpine.

Union Peak, August 12, 1894 (No. 897).

Artemisia scopulorum, Gray, Proc. Acad. Phila. 66 (1863).

Probably in all our alpine regions.

Union Peak, August 13, 1894 (No. 989); La Plata Mines, August 22, 1895 (No. 1779).

Artemisia tridentata, Nutt. Trans. Am. Phil. Soc. (II) vii, 398 (1841).

This is the shrub that is generally designated by the term "sage brush," whereas the term sage is applied indiscriminately to the preceding. It is, perhaps, the most characteristic shrub of the Wyoming plains and valleys. Its presence indicates soil of good quality, reasonably free from alkali. Of very slow growth, but on some creeks reaching the dignity of small trees and furnishing excellent fuel.

Boulder Creek, August 25, 1894 (No. 1111).

Artemisia trifida, Nutt. l. c.

No specimens of this species are at hand but it is known to be in the state.

Noted by Prof. W. C. Knight, on Seminoe Mountains, May 6, 1896, at about 8,000 ft. altitude.

Petasites sagittata, Gray, in Brew. & Wats. Bot. Cal. i, 407 (1876).
Tussilago sagittata, Pursh.

Rare; in a wet, boggy meadow.

Pole Creek, May 25, 1894 (No. 81).

Arnica alpina, Olin. Mon. Arn. Upsala (1799).

Very abundant about Table Mountain, where it was collected June 2, 1894 (No. 148); June 30, 1895 (No. 1383).

Arnica amplexicaulis, Nutt. Trans. Am. Phil. Soc. vii, 408.

Frequent in copses in fertile subalpine valleys.

Teton Mountains, August 21, 1894 (No. 933); Centennial Hills, August 19, 1895 (No. 1702).

Arnica Chamissonis, Less. Linnæa, vi, 317 (1831).

Frequent in the mountains.

Pine Creek, by B. C. Bufum; Union Peak, August 13, 1894 (No. 995); La Plata Mines, August 23, 1895 (No. 1785).

Arnica cordifolia, Hook. Fl. Bor. Am. i, 331 (1833).

Exceedingly abundant; in woods and copses in the mountains from their bases to their summits. Somewhat variable, the radical leaves often reduced and ovate rather than cordate.

Horse Creek June 9, 1894 (No. 215); Union Pass, August 10, 1894 (No. 871); also from the Centennial Valley.

Arnica foliosa, Nutt. Trans. Am. Phil. Soc. vii, 407.

Probably rare; collected only on upper Wind River, August 10, 1894 (No. 766).

Arnica foliosa incana, Gray, Bot. Cal. i, 416 (1876).

Specimens from Pine Creek, by Prof. B. C. Buffum, presumably nearly typical; others less so from Saratoga, July 2, 1893, by J. D. Parker.

Arnica latifolia, Bong. Veg. Sitch. 147.
    A beautiful subalpine species, probably in all of our mountains. Bald Mountain, B. C. Buffum, in August 1892; Union Pass, August 11, 1894 (No. 836); noted in Medicine Bow Mountains.

Arnica longifolia, Eaton, Bot. King Exp. 186.
    Decidedly rare; Teton Mountains, August 21, 1894 (No. 957).

Senecio amplectens taraxacoides, Gray, Proc. Acad. Phila. 67 (1863).
    Very rare; only a few specimens from the Teton Mountains, August 22, 1894 (No. 987).

Senecio aureus, L. Sp. Pl. ii, 870 (1753).
    This and many of its varieties are well represented in our flora. The individual variation is often so great that it is difficult to assign some specimens to any of the numerous varieties already created.
    Specimens from Green River, August 26, 1894 (No. 1036), and Wind River, August 10, 1894 (No. 760) I think are nearly typical, as are also some from Beaver Creek and Laramie, by B. C. Buffum.

Senecio aureus borealis, T. & G. Fl. N. A. ii, 442.
    On the rocky slopes of the Teton Mountains, August 21, 1894 (No. 979).

Senecio aureus croceus, Gray, Proc. Acad. Phila. 68 (1863).
    This is a frequent form in subalpine meadows.
    Union Pass, August 10, 1894 (No. 858); La Plata Mines, August 20, 1895 (No. 1753).

Senecio aureus obovatus, T. & G. Fl. N. A. ii, 442.
    Typical specimens from Big Wind River, August 5, 1894 (No. 704); doubtful ones from La Plata Mines, August 23, 1895 (No. 1769).

Senecio Balsamitæ (?) Muhl. Willd. Sp. Pl. iii, 1998 (1804).
    I think there is little doubt of the two sets of specimens before me belonging here; they lack the root leaves upon which the description partly hinges, but seem normal otherwise.
    Eagle Rock Canon, August 2, 1892, B. C. Buffum; Cummins, July 30, 1895 (No. 1492).

Senecio canus, Hook. Fl. Bor. Am. i, 333 (1834).
    This strongly marked species is of frequent occurrence in the state. Our specimens are large, stoloniferous and slightly decumbent at base.

Union Pass, August 10, 1894 (No. 761); Table Mountain, June 30, 1895 (No. 1364).

Senecio crassulus, Gray, Proc. Am. Acad. xix, 54.

Probably in the subalpine regions of all our mountains.

A large form in Union Pass, August 10, 1894 (No. 809); smaller and more typical, Union Peak, August 13, (No. 1027); La Plata Mines, August 21, 1895 (No. 1770).

Senecio Douglasii, D C. Prodr. vi, 429 (1837).

This is of frequent occurrence, the three main forms being represented : 1. Glabrous, stems branched, leaves broad, with broad, unequal lobes ; rays long, narrow (6-16). This is probably the suppressed *Senecio fastigiatus*, Gray. 2. Slightly pubescent-tomentose stems paniculately branched only at summit ; leaves pinnately parted into linear lobes, the rays few and inconspicuous. This form is Bentham's *Senecio longilobus*. 3. Glabrous, stems very numerous from a woody base, strict ; leaves entire, linear ; inflorescence a short cymose panicle ; rays long and conspicuous. This is *Senecio Riddellii*, T. & G.

I see no reason why forms so distinct should not be kept separate. 1. From Pole Creek, July 1, 1895 (No. 1389); 2. Sybille Hills, July 8, 1894 (No. 302), and Cummins, July 27, 1895 (No. 1441); 3. Cummins, July 27, 1895 (No. 1440).

Senecio eremophilus, Richards. App. Frankl. Journ. Ed. 2, 31.

Infrequent ; Cummins, July 29, 1895 (No. 1491).

Senecio Fendleri, Gray, Pl. Fendl. 108.

A very common form in stony clay hills.

Pole Creek, June 2, 1894 (No. 124); Centennial Valley, June 9, 1895 (No. 1297).

Senecio hydrophilus, Nutt.

Frequent, but never plentiful ; in wet and sometimes boggy places.

Bacon Creek, August 15, 1894 (No. 915); Cummins July 28, 1895 (No. 1458).

Senecio integerrimus, Nutt. Gen. ii, 165.

Very frequent in fertile, open valleys in the mountains.

Union Pass, August 14, 1894 (No. 891); Teton Mountains, August 21, 1894 (No. 1002); Table Mountain, June 26, 1895 (No. 1333).

Senecio lugens, Richards. Bot. App. 748 (1823).

Very common and of many forms which probably are worthy of varietal distinction. The following numbers appear early on hillsides and valleys and are probably near the type:

Table Mountain, June 2, 1894 (No. 128); Centennial Valley, June 9, 1895 (No. 1305).

Senecio lugens foliosus, Gray, Bot. Cal. i, 413.

On hillsides, in rich loam soil among undershrubs.

Union Pass, August 13, 1894 (No. 999); Centennial Valley, June 9, 1895 (No. 1246).

Senecio lugens megalocephalus, n. var.

Radical leaves large, irregularly and coarsely dentate; cauline leaves small or wanting; stem branched from near the base into several slender branches 8-12 inches long, each bearing a few (3-5) large heads (5-8 lines high). The whole plant is very tardily glabrate. In thickets on moist hillsides.

Centennial Valley, June 9, 1895 (No. 1252).

Senecio rapifolius, Nutt.

Said to be a very rare plant.

Sweetwater River, September, 1894 (No. 1180); Laramie Peak, August 7, 1895 (No. 1589).

Senecio serra, Hook. Fl. Bor. Am. i, 333 (1834).

Infrequent; near Warm Spring Creek, August 10, 1894 (No. 771).

Senecio serra integriusculus, Gray, Syn. Fl. 387 (1884). *S. Andinus,* Nutt.

A rare plant; in the Union Pass hills, August 10, 1894 (No. 876).

Senecio triangularis, Hook. Fl. Bor. Am. i, 332 (1834).

On wet banks in copses in mountain canons.

Teton Mountains, August 21, 1894 (No. 936); Cummins, July 31, 1895 (No. 1519).

Senecio werneræfolius, Gray, Proc. Am. Acad. xix, 54.

Frequent in the "red hills" (Tertiary) about Laramie, June 15, 1894 (No. 224); Table Mountain, June 29, 1895 (No. 1379).

Tetradymia canescens, D C. Prodr. vi, 440 (1837).

On the dry plains of the Platte; Big Muddy Creek, July 23, 1894 (No. 605).

Tetradymia canescens inermis, Gray.

While not separated from the species by any strong botanical characters, yet readily distinguished by foliage and general habit.

Wallace Creek, July 29, 1894 (No. 656).

Cnicus altissimus filipendulus, Gray, Proc. Am. Acad. xix, 56.

Specimens received from Sheridan Experiment Farm seem to belong to this variety, probably introduced with farm seeds. September, 1895.

Cnicus Americanus, Gray, Proc. Am. Acad. xix, 56.

In the dry foothills near the Laramie River; not frequent.

Cummins, July 29, 1895 (No. 1512).

Cnicus arvensis, Hoffm. Deutch. Fl. Ed. 2, I: Part 2, 130 (1804). *Carduus arvensis,* (L.)Robs.

Thus far this, the *Canada Thistle,* has been reported from only one place in the state.

Specimens from Sheridan, September 1895, J. F. Lewis.

Cnicus Drummondii, Gray, Proc. Am. Acad. x, 40. *Carduus pumilus,* Hook.

Very frequent throughout the state.

Laramie, September, 1893; Union Pass, August 14, 1894 (No. 881); Cummins, July 27, 1895 (No. 1469).

Cnicus lanceolatus, Hoffm. *Carduus lanceolatus,* L.

Becoming plentiful in vacant lots and along the irrigation ditches.

Cnicus ochrocentrus, Gray, Proc. Am. Acad. xix, 57 (1883).

In dry, open ground in the hills; frequent.

Laramie Hills, July 7, 1894 (No. 414); Gros Ventre River, August 22, 1894 (No. 1070).

Cnicus undulatus, Gray, Proc. Am. Acad. x, 42 (1874). *Carduus undulatus,* Nutt.

Frequent; Cheyenne, August 11, 1891, B. C. Buffum; Laramie, October 2, 1894 (No. 1156).

Cnicus Virginianus, Pursh, Fl. ii, 506 (1814). *Carduus Virginianus,* L.

Very rare; a few specimens from a dry bank, Cummins, July 30, 1895 (No. 1563).

Crepis acuminata, Nutt. Trans. Am. Phil. Soc. vii, 437.

On dry, sometimes stony ground, in the hills and mountains.

Laramie, July 7, 1894 (No. 356); Union Pass, August 10, 1894 (No. 919).

Crepis elegans, Hook. Fl. i, 297.
>A very rare plant; noted but once.
>Union Pass, August 12, 1894 (No. 1076).

Crepis glauca, T. & G. Fl. N. A. ii, 488 (1843).
>Frequent in wet alkali meadow lands.
>Meadow Creek, August 9, 1894 (No. 787); Centennial Valley, August 16, 1895 (No. 1673).

Crepis intermedia, Gray, Syn. Fl. 432 (1884).
>Rare; secured by B. C. Buffum on the Wind River, August 2, 1892.

Crepis intermedia gracilis, Gray, Syn. Fl. 432 (1884).
>Frequent on dry hillsides in the Laramie range.
>Table Mountain, June 30, 1895 (No. 1393).

Crepis runcinata, T. & G. Fl. N. A. ii, 487 (1843).
>Infrequent; secured by B. C. Buffum near Big Wind River, August 1, 1892.

Hieracium albiflorum, Hook. Fl. i, 298.
>Abundant on a dry hillside among the fallen timber in a burned-over district.
>Centennial Hills, August 16, 1895 (No. 1678).

Hieracium Canadense, Michx. Fl. ii, 86 (1803).
>Infrequent; Wolf Creek, August 18, 1892, B. C. Buffum.

Hieracium Fendleri, Schultz, Bip. Bonplandia, ix, 173.
>As near as I can judge without other material for comparison, these specimens are nearly typical.
>Plentiful on the banks of the Little Laramie River, in the Centennial Valley, August 25, 1895 (No. 1857).

Hieracium gracile, Hook. Fl. i, 298.
>Very typical specimens from the Medicine Bow Mountains, where in the alpine region it is found in the greatest profusion. August 23, 1895 (No. 1802).

>A very diminutive form was secured on the Teton Mountains; radical leaves only 3-6, stems single, bearing one to four heads. Var. **minimum** would be a suitable designation. August 21, 1894 (No. 1060).

Hieracium Scouleri, Hook. Fl. i, 298.
Very abundant on the Gros Ventre Hills, but not observed east of the Wind River Mountains. August 23, 1894 (No. 963).

Troximon aurantiacum, Hook. Fl. i, 300, t. 104.
Infrequent; Pole Creek, June 29, 1895 (No. 1342).

Troximon aurantiacum purpureum, Gray, Proc. Am. Acad. xix, 72.
Found occasionally in the vicinity of Laramie; Pole Creek, June 30, 1895 (No. 1376); Hutton's Grove, August 9, 1891, B. C. Buffum.

Troximon cuspidatum, Pursh, Fl. Am. Sept. 742 (1814). *Nothocalais cuspidata* (Pursh) Greene.
Somewhat variable; only a few of my numerous specimens typical; possibly some of them should be cut out. Frequent but scattering. Laramie, August 1893; Wallace Creek, July 29, 1894 (No. 673); from several other places and noted in many localities.

Troximon glaucum, Pursh, Fl. 505 (1814). *Agoseris glauca* (Pursh) Greene.
Frequent in the grassy valleys in the hills and mountains.
Eagle Rock Canon, August 22, 1892, B. C. Buffum; Union Pass, August 10, 1894 (No. 870).

Troximon glaucum dasycephalum, T. & G. Syn. Fl. 432 (1884).
Seemingly nearly alpine and occasionally very abundant.
Union Pass, August 10, 1894 (No. 868); La Plata Mines, August 21, 1895 (No. 1765).

Troximon glaucum laciniatum, Gray, Bot. Cal.
This form I have never been able to locate to my satisfaction, but I am unable to place it under any other name. It is of very frequent occurrence among the sage brush in the Laramie Hills; June 2, 1894 (No. 125); Pole Creek, June 29, 1895 (No. 1376).

Taraxacum officinale, Weber, Prim. Fl. Holst. 56 (1780). *T. Taraxacum*, (L.) Karst.
Apparently the *Dandelion* found its ideal home when it reached Laramie. It occupies every foot of ground along the irrigation ditches of our streets and takes complete possession of the lawns where eternal warfare is not waged upon it. In luxuriant growth and blossom from April to November (No. 80).

Taraxacum officinale alpinum, Koch. Fl. Germ. & Helv. 428 (1837).
Of the several forms of native Dandelions this seems to be the most frequent. Abundant in the high, grassy valleys of the Laramie range.
Pole Creek, June 2, 1894 (No. 109); Centennial Valley, August 17, 1895 (No. 1715).

Taraxacum officinale lividum, (?) Koch.
Some specimens collected by B. C. Buffum, 1892, are doubtfully placed here, and some with nearly entire glaucescent leaves are for the present passed over.

Lactuca leucophæa, Gray, Proc. Am. Acad. xix, 73.
Probably very rare in the state.
Centennial Valley, August 16, 1895 (No. 1674).

Lactuca Ludoviciana, DC. Prodr. vii, 141 (1838).
Rare; possibly confined to the eastern part of the state.
Laramie Peak, August 7, 1895 (No. 1596).

Lactuca pulchella, DC. Prodr. vii, 134 (1838).
Very frequent in the fertile soil of valleys.
Blue Grass Creek, July 13, 1894 (No. 445); Laramie, September 16, 1894 (No. 1146).

Sonchus asper, (L.) All. Fl. Ped. i, 222 (1785).
This has found its way into several parts of the state; Lander, August 3, 1894 (No. 872); Evanston, September 1, 1894, sent by Dr. Solier.

Lygodesmia grandiflora, T. & G. Fl. ii, 485.
Very rare; some specimens by B. C. Buffum from near Tie Siding, July 18, 1891.

Lygodesmia juncea, Don. Edinb. Phil. Journ. vi, 311 (1829).
A "weedy" plant very common on the Laramie Plains.
University campus, September 26, 1894 (No. 1165). *Skeleton Weed.*

Stephanomeria minor, Nutt. Trans. Am. Phil. Soc. (II) vii, 427 (1841).
Ptiloria tenuifolia, (Torr.) Raf.
Probably not rare, though observed but twice.
Gros Ventre Hills, August 16, 1894 (No. 925); also a small form of it from Laramie Peak, August 24, 1895 (No. 1624).

Stephanomeria runcinata, Nutt. l. c. 128. *Ptiloria pauciflora,* (Torr.) Raf.

Infrequent; on a ridge of disintegrated rock, Garfield Peak, July 27, 1894 (No. 655).

CAMPANULACEÆ.

Campanula Parryi, Gray, Syn. Fl. Suppl. 395.

Rare; probably confined to the high, grassy valleys of the southern part of the state.

Cummins, July 31, 1894 (No. 1495).

Campanula rotundifolia, L. Sp. Pl. 163 (1753).

One of the very commonest plants in all our hills and mountains, at least where a reasonable amount of moisture is to be found.

Cold Springs, July 14, 1894 (No. 448); Union Pass, August 11, 1894 (No. 814); Cummins, August 1, 1895 (No. 1540). *Blue Bell.*

Specularia perfoliata, A. DC. Mon. Camp. 351 (1830). *Legouzia perfoliata,* (L.) Britton.

Not infrequent in the eastern part of the state.

Whalen Canon, July 18, 1894 (No. 514); Laramie Peak, August 8, 1895 (No. 1657).

ERICACEÆ.

Vaccinum cæspitosum, Michx. Fl. Bor. Am. i, 234 (1803).

Very abundant in the Medicine Bow Mountains and producing the small, sweet berries in profusion.

Centennial Hills, August 18, 1895 (No. 1728). *Blueberry.*

Vaccinum Myrtillus, L. Schk. Handb. t. 107.

In openings in the Spruce timber at high elevations; infrequent. July 1892, by B. C. Buffum; Centennial Hills, June 9, 1895 (No. 1292).

Arctostaphylos Uva-ursi, (L.) Spreng. Syst. ii, 287 (1825).

Throughout the state, both on open and on wooded hillsides. Laramie Hills, May 18, 1894 (No. 1214). *Kinnikinick.*

Bryanthus empetriformis, Gray, Proc. Am. Acad. vii, 377.

A beautiful alpine species; in small basin-like heath among the Spruce trees.

Union Peak, August 13, 1894 (No. 1006).

Kalmia glauca, Ait. Hort. Kew. ii, 64 t. 8 (1811).
   Rare; on the margin of a little lake at 9,000 ft., Teton Mountains, August 21, 1894 (No. 952).

Pyrola chlorantha. Sw. Act. Holm. 1810, t. 5 (1810).
   In deep, shaded ravines; Cummins, July 31, 1895 (No. 1505).

Pyrola minor, L. Sp. Pl. 396 (1753).
   On the shaded border of a lake immediately at the base of one of the perpetual snow banks of the Medicine Bow Mountains.
   La Plata Mines, August 23, 1895 (No. 1825).

Pyrola rotundifolia, L. Sp. Pl. 396 (1753).
   Infrequent; Teton Mountains, August 21, 1894 (No. 947).

Pyrola rotundifolia uliginosa, Gray, Man. Ed. 2, 259 (1856). *P. uliginosa,* Torr.
   Much more frequent than the species; abundant in cold, shaded, wet places in the mountains.
   Centennial Hills, August 19, 1895 (No. 1729).

Pyrola secunda, L. Sp. Pl. 396 (1753).
   Very abundant in deep woods, especially of high, rich valleys.
   Bald Mountain, August 15, 1892; Union Pass, August 10, 1894 (No. 802); Cummins, July 30, 1895 (No. 1504).

Moneses uniflora, Gray. 273 (1848).
   In the densely shaded woods about a mountain lake on the Tetons, August 21, 1894 (No. 945); also by B. C. Buffum in a gulch near Bald Mountain, 1892.

Chimaphila umbellata, Nutt. Gen. i. 274 (1818).
   On mountain sides in the woods or in the shade of overhanging cliffs.
   Teton Mountains, August 21, 1894 (No. 946); Laramie Peak, August 6, 1895 (No. 1616).

MONOTROPACEÆ.

Pterospora Andromedea, Nutt. Gen. i, 269 (1818).
   In the pine woods, frequent but not abundant.
   Snake River, August 22, 1894 (No. 985); Centennial Valley, August 17, 1895 (No. 1684).

Monotropa Hypopitys, L. Sp. Pl. 387 (1753). *Hypopitys Hypopitys,* (L.) Small.
   Infrequent; Centennial Hills, August 17, 1895 (No. 1708).

PRIMULACEÆ.

Primula farinosa, L. Sp. Pl. 143 (1753).
Very rare; not found by the writer, but three fine specimens received from Mr. Houghton, who collected them on the Little Laramie River, June 8, 1894 (No. 187).

Primula Parryi, Gray, Am. Journ. Sci. II, xxxiv, 257.
A very rare plant; collected by B. C. Buffum in the Wind River Mountains, July 21, 1892.

Androsace filiformis, Retz. Obs. ii, 10.
Very abundant in wet, caved in places on mountain streams.
Pole Creek, June 27, 1895 (No. 1318); La Plata Mines, August 23, 1895 (No. 1848).

Androsace occidentalis, Pursh, Fl. Am. Sept. 137 (1814).
Common on dry creek banks, especially in cattle wallows where other vegetation has been killed.
Horse Creek, June 9, 1894 (No. 194); Centennial Valley, June 8, 1895 (No. 1244).

Androsace septentrionalis, Lam. Ill. t. 98, f. 2.
On moist hillsides at both the higher and the lower altitudes.
Union Pass, August 13, 1894 (No. 1030); Table Mountain, June 27, 1895 (No. 1332).

Androsace septentrionalis subumbellata, n. var.
A diminutive alpine form may receive this name. Plant only ½ to 1 inch high; leaves mostly entire; scapes few and one flowered, or, if umbelliferous only three to five; calyx shorter than the corolla, its lobes noticeably shorter than its tube.
On a grassy hillside near the summit of Union Peak, August 13, 1894 (No. 998).

Dodecatheon Meadia, L. Sp. Pl. 144 (1753).
On the partly shaded banks of small brooklets in the Laramie Hills.
Pole Creek, June 27, 1895 (No. 1329).

Dodecatheon pauciflorum, Greene.
The wet meadows in the Laramie River bottoms, in some places, are so densely covered with this plant that at a distance they suggest immense lakes of purplish-blue water. June 19, 1894 (No. 268); Centennial Valley, June 9, 1895 (No. 1312).

Dodecatheon, sp.
    Some very small specimens secured by B. C. Buffum at Bald Mountain, with single flowered scapes, probably belong to the var. *alpina*, Gray, of *D. Meadia*. August 8, 1892.

Steironema ciliatum, Baudo, Ann. Sci. Nat. (II) xx, 346 (1843).
    Rich soil in thickets on streams.
    Laramie Peak, August 7, 1895 (No. 1582); Bald Mountain, August, 1892.

Glaux maritima, L. Sp. Pl. 148 (1753).
    On the low, wet shores of alkali ponds and lakes; abundant.
    Laramie, June 16, 1894 (No. 248).

OLEACEÆ.

Fraxinus viridis, Michx. f. Hist. Arb. 3: 115, t. 10 (1813). *F. lanceolata*, Borck.
    Rare in the parts of the state collected.
    Platte River, July 14, 1894 (No. 480).

APOCYNACEÆ.

Apocynum androsæmifolium, L. Sp. Pl. 213 (1753).
    Frequent on grassy slopes in the foothills of the Platte.
    Whalen canon, July 19, 1894 (No. 535); Laramie Peak, August 7, 1895 (No. 1632).

Apocynum cannabinum, L. Sp. Pl. 213 (1753).
    Rare in the districts collected, but probably frequent in the northeast.
    On the banks of the Platte, July 11, 1894 (No. 396); also on the Big Muddy, July 23, 1894 (No. 596).

ASCLEPIADACEÆ.

Acerates viridiflora linearis, Gray, Syn. Fl. 2.
    Rare; occasional specimens on the banks of the Platte River.
    Fairbanks, July 14, 1894 (No. 489).

Asclepias cryptoceras, Watson, King's Rep. (1871).
    This is a rare plant, coming into our range only in the western part.
    Snake River, May 29, 1892, by Fred McCoullough.

Asclepias Hallii, Gray, Proc. Am. Acad. viii, 69.
Infrequent; on the banks of a clayey ravine at Wood's Landing, July 31, 1895 (No. 1556).

Asclepias speciosa, Torr. Ann. Lyc. N. Y. ii, 218 (1827).
Frequent in the eastern part of the state.
Lusk, July 21, 1894 (No. 575); McGill's Ranch, on the Laramie River, August 3, 1895 (No. 1573).

Asclepias verticillata pumila, Gray, Proc. Am. Acad. xii, 71 (1876).
Rare; in a sandy canon leading to the Platte; Fairbanks, July 14, 1894 (No. 466).

GENTIANACEÆ.

Gentiana affinis, Griseb. Hook. Fl. Bor. Am. i, 56 (1834).
In wet meadows at 8,000 to 10,000 ft.
Upper Wind River, August 8, 1894 (No. 754); Snake River, August 22, 1894 (No. 960); La Plata Mines, August 22, 1895 (No. 1853).

Gentiana Amarella acuta, Hook. *G. acuta,* Michx.
In meadow lands on the Laramie River and its tributaries; very abundant.
Cummins, July 30, 1895 (No. 1543); Centennial Valley, August 25, 1895 (No. 1852).

Gentiana Amarella stricta, Watson, King's Rep. (1871).
Rare; specimens from Wolf Creek, August, 1892, by B. C. Buffum.

Gentiana calycosa, Griseb. in Hook. Fl. Bor. Am. ii, 68 (1838).
This beautiful alpine species probably in all of our higher mountains.
Tetons, August 21, 1894 (No. 1057); La Plata mines, August 22, 1895 (No. 1767).

Gentiana frigida, Hænke. Jacq. Coll. ii, 13.
As rare as it is odd; on the shore of a small alpine lake, La Plata Mines, August 23, 1895 (No. 1804).

Gentiana humilis, Stev. Act. Mosq. iii, 258.
Frequent on the banks of small creeks in the hills and mountains.
Horse Creek, June 9, 1894 (No. 220); Centennial Valley, June 9, 1895 (No. 1313).

Gentiana Oregana, Engelm. Gray, Syn. Fl. ii, pt. i, 122 (1886).
   Infrequent; in a mountain valley, at 8,000 ft.; Cummins, July, 29, 1895 (No. 1527).

Gentiana Parryi, Engelm. Trans. St. Louis Acad. ii, 218, t. 10.
   Wet, subalpine valleys.
   Cummins, July 29, 1895 (No. 1526); Bald Mountain, August 17, 1892.

Gentiana serrata holopetala, Gray, Bot. Cal. i, 481 (1876).
   Under this name I have two quite distinct forms, neither of which is typical, but lacking salient characters enough to separate them. The one is small and simple stemmed, the other freely branched from the base. Both with unusually large showy flowers conspicuously fringed. Remarkably abundant in high, wet valleys.
   Union Pass, August 10, 1894 (No. 865); Cummins, July 30, 1895 (No. 1539); Centennial Valley, August 17, 1895 (No. 1725); La Plata, August 21, 1895 (No. 1766).

Gentiana tenella, Rottb. Act. Haffn. x, 436, t. 2.
   Rare; Cummins, July 27, 1895 (No. 1518).

Pleurogyne rotata, Griseb. Gent. 309 (1839).
   Very rare; on the hummocks in a swampy meadow, Centennial Valley, August 18, 1895 (No. 1701).

Swertia perennis, L. Sp. Pl. i, 226 (1753).
   In subalpine meadow swamps; frequent.
   Union Pass, August 13, 1894 (No. 988); Centennial Hills, August 20, 1895 (No. 1742).

Frasera speciosa, Dougl. Hook. Fl. Bor. Am. ii, 66 (1838).
   By reason of its size, a conspicuous and characteristic object both on the timbered and denuded mountain slopes.
   Wind River Mountains, August 12, 1894 (No. 935); Laramie Peak, August 6, 1895 (No. 1609).

POLEMONIACEÆ.

Phlox bryoides, Nutt. Journ. Acad. Phila. II, i, 153 (1848).
   This is very abundant on rather naked limestone ridges where they crop out on the Laramie Plains and in the foothills. May 25, 1894 (No. 67); June 7, 1895 (No. 1280).

Phlox cæspitosa, Nutt. Journ. Acad. Phila. vii, 41 (1834).
   One of the earliest flowers on the plains; very abundant; pink flowered forms are common.
   Laramie, May 12, 1894 (No. 8), and at various other times.

Phlox Douglasii, Hook. Fl. Bor. Am. 2 : 73, t. 158 (1834).
   On the sides of dry, sandy ravines and ridges; infrequent.
   Uva, July 10, 1894 (No. 397); also from Beaver Divide, July, 1891.

Phlox longifolia, Nutt. Journ. Acad. Phila. vii, 41 (1834).
   Leaves mostly less than an inch in length, otherwise the plant seems typical; very abundant in the Laramie Hills; June 7, 1894 (No. 182), and at several other times.

Phlox longifolia brevifolia, Gray.
   Specimens perfectly in accord with description in Syn. Fl. were obtained at La Plata Mines, August 24, 1894 (No. 1827).

Phlox nana, Nutt. Pl. Gamb. 153.
   If this determination is right, which it is if the descriptions go for anything, this extends the range of this plant quite a little.
   Collected on Snake River, May 29, 1892, by Fred McCoullough.

Gilia aggregata, Spreng. Syst. i. 626 (1825).
   Widely distributed, but never in any great abundance; Beaver Creek, July 17, 1892; Bell Springs, July 4, 1891; Snake River, August 22, 1894 (No. 961).

Gilia aggregata attenuata, Gray, Syn. Fl. II, i, 145 (1886).
   On an abrupt, stony, gravelly bank of the Laramie River. Cummins, July 29, 1895 (No. 1522).

Gilia Breweri, Gray, Proc. Am. Acad. viii, 266.
   Infrequent; possibly confined to the Pacific slope.
   Gros Ventre River, August 18, 1894 (No. 1094).

Gilia gracilis, Hook. Bot. Mag. t. 2924 (1829).
   Very abundant on the rich sandy loam of the small valleys of the Laramie Hills and in similar locations throughout the state.
   Pole Creek, June 2, 1894 (No. 107); Inyan Kara Divide, August 30, 1892.

Gilia inconspicua, Sweet, Hort. Brit. 286 (1826).
   Thus far only from the Big Wind River, August 3, 1892, by B. C. Buffum; August 5, 1894 (No. 709).

Gilia inconspicua sinuata, Gray, Proc. Am. Acad. viii, 278.

This extends the range of this plant northward somewhat ; in the sandy bed of a dry creek ; Muskrat July 30, 1894 (No. 683).

Gilia linearis, Gray, Proc. Am. Acad. xvii, 223 (1882). *Collomia linearis,* Nutt.

This is by far the most frequent of our Gilias, common in the rich loam along streams everywhere.

Pole Creek, June 2, 1894 (No. 108); Sybille Creek, July 8, 1894 (No. 401).

Gilia minima, Gray, Proc. Am. Acad. viii, 269 (1880). *Navarretia minima,* Nutt.

Not found on the Atlantic slope.

Silver Creek, August 26, 1894 (No. 1129).

Gilia nudicaulis, Gray, Proc. Am. Acad. viii, 266 (1870).

The type locality is in the western part of this state, and the plant is probably common throughout the state, but so small and inconspicuous as usually to be passed over.

Horse Creek, June 9, 1894 (No. 193).

Gilia Nuttallii, Gray, Proc. Am. Acad. viii, 267 (1870)

An undoubted specimen of this species without data, by B. C. Buffum, in 1892. Probably near Bald Mountain, August, 1892.

Gilia spicata, Nutt. Journ. Acad. Phil. (II) i, 156 (1848).

Frequent about the sand dunes in the foothills.

Pass Creek, June 20, 1892 ; Laramie Hills, June 7, 1894 (No. 175). My number 259 approaches *G. spicata capitata,* Gray, pretty closely, but I do not think it typical.

Polemonium confertum, Gray, Proc. Acad. Phila. 1863, 63 (1863).

A beautiful alpine plant on grassy or rocky slopes.

Union Peak, August 13, 1894 (No. 991); La Plata Mines, August 22, 1895 (No. 1823).

Polemonium confertum mellitum, Gray, l. c.

In crevices and on ledges on the higher rocky summits of the Laramie Mountains.

Richardson's Peak, June 9, 1894 (No. 208); Laramie Peak, August 7, 1895 (No. 1625).

Polemonium humile pulchellum, Gray, Syn. Fl. II. i, 150 (1886).

What seems to belong here was secured by B. C. Buffum in 1892 ; no other data.

Polemonium occidentale, Greene. Pittonia ii, 75 (1890).
    Name communicated by Prof. Greene. Rare; in spring bog on Muddy Creek, August 25, 1894 (No. 1104).

HYDROPHYLLACEÆ.

Hydrophyllum occidentale, Gray, Proc. Am. Acad. x, 314.
    This, I am sorry to say, I have distributed as *H. Virginicum,* — a piece of carelessness, for the specimens plainly enough belong here.
    Very abundant in the copses on our streams.
    Pole Creek, June 2, 1894 (No. 89); Table Mountain, June 29, 1895 (No. 1408).

Ellisia Nyctelea, L. Sp. Pl. Ed. 2, 1662 (1763). *Macrocalyx Nyctelea,* Kuntze.
    On dry loam soil on creek banks; observed only on the east slopes of the Laramie range.
    Table Mountain, June 27, 1895 (No. 1350).

Phacelia circinata, Jacq. Eclog. 135 t. 91.
    Of the synonomy and citations applicable to my specimens, I am in doubt, but all answer to the above in Syn. Fl. II. i, 159.
    Inyan Kara Divide, August 30, 1892; Union Peak, August 14, 1894 (No. 1082); Pole Creek, June 26, 1895 (No. 1323).

Phacelia Franklinii, Gray, Man. Ed. 2, 329 (1856).
    This gives one more locality for the range of this plant.
    Bacon Creek, August 15, 1894 (No. 914).

Phacelia glandulosa, Nutt. Fl. Gambl. 160.
    Infrequent and scattering; Pole Creek, June 30, 1895 (No. 1361); also noted at Cummins.

Phacelia Menziesii, Torr. Watson, King Rep. 252.
    These handsome specimens were sent to us from Snake River, May 29, 1892. Fred McCoullough.

Phacelia sericea, Gray. Am. Journ. Sci. ser. 2, (1862) xxxiv, 254.
    Very common in moist, partly shaded ground in the hills and along streams.
    Pole Creek, June, 2, 1894 (No. 102); a white flowered form from the Centennial Valley, June 9, 1895 (No. 1283).

Phacelia sp.

A small annual having affinities with both *P. circinata*, and *P. Franklinii*, must for the present be passed over.

Sybille, July 8, 1894 (No. 318).

BORAGINACEÆ.

Coldenia Nuttallii, Hook. Kew. Journ. Bot. iii, 296 (1851).

In a sandy canon or valley leading to the Big Wind River, August 6, 1894 (No. 719); observed in no other locality.

Echinospermum floribundum, Lehm in. Hook. Fl. 2: 84, t. 164 (1834). *Lapula floribunda*, Greene.

Growing almost at the water's edge on some of the streams in the eastern part of the state.

Sybille Creek, July 7, 1894 (No. 348); Pole Creek, June 29, 1895 (No. 1367).

Echinospermum Lapula, Lehm. Asperif. 121 (1818). *Lapula Lapula*, (L.) Karst.

This weed seems to have found its way into waste places about town.

Laramie, June 17, 1891; also at Uva, July 10, 1894 (No. 424).

Echinospermum Redowskii occidentale, Watson, King's Exp. 246 (1871). *Lapula Texana*, (Scheele) Britton.

Common everywhere; Laramie, June 28, 1894 (No. 294); Blue Grass Hills, July 8, 1894 (No. 305).

Echinospermum Redowskii cupulatum, Gray, in Brew. & Wats. Bot. Cal. i, 530 (1876).

This, I note, has been reduced to the same as the preceding, but to say the least there is a marked difference between the nutlets.

By B. C. Buffum, in the northeastern part of the state, August 19, 1892.

Allocarya Nelsonii, Greene, Erythea iii, 48 (1895).

The original description is as follows: "Annual, diffuse, the stoutish and somewhat succulent branches strigose-pubescent, six inches long, rather densely racemose throughout and with a short bract subtending each pedicel; nutlets ¼ line long, ovate-lanceolate, carinate ventrally almost down to the nearly basal rounded or obscurely trigonous scar, the back with rather few and sharp transverse ridges beset with tufts of uncinate-tipped bristles, the intervals with low, muriculate-roughened tuberculations."

Only one small patch of these plants observed; they had taken possession of a few depressions, possibly old buffalo wallows, on bottom land near Silver Creek, August 26, 1894 (No. 1198).

Allocarya scopulorum, Greene, Pitt. i, 16 (1887).

Rather infrequent; Upper Wind River, August 10, 1894 (No. 768); Centennial Valley, August 25, 1895 (No. 1850).

Krynitzkia Californica subglochidiata, Gray, Bot. Cal. i, 526.

Common on dry loam soil near streams.

Cooper Creek, June 19, 1892; Table Mountain June 28, 1895 (No. 1311).

Krynitzkia crassisepala, Gray, Proc. Am. Acad. xx, 268 (1885). *Cryptanthe crassisepala,* (T. & G.) Greene.

Frequent on sandy plains.

Laramie, August 8, 1891; Blue Grass Creek, July 8, 1894 (No. 304).

Krynitzkia Fendleri, Gray, l. c. *Cryptanthe Fendleri,* (Gray) Greene.

Infrequent; Cummins, July 25, 1895 (No. 1523).

Krynitzkia fulvocanescens, Gray, Proc. Am. Acad. xx, 280 (1885). *Oreocarya fulvocanescens,* (Gray) Greene.

Fine specimens of this were secured by Prof. Buffum, at Cooper Creek, June 18, 1892.

Krynitzkia glomerata, Gray, Proc. Am. Acad. xx, 279 (1885). *Oreocarya glomerata,* (Gray) Greene.

This is perhaps our commonest Krynitzkia; frequent on dry hillsides, railroad embankments and on the plains.

Laramie, July 8, 1894 (No. 418); Table mountain, June 30, 1895 (No. 1362); a very large, coarse form from Uva, July 10, 1894 (No. 388).

Krynitzkia glomeriflora, Greene.

Frequent in dry, rich loam soil along streams.

Pole Creek, near Table Mountain, June 2, 1894 (No. 152); Centennial Valley, June 9, 1895 (No. 1335).

Krynitzkia Jamesii, Gray, Proc Am. Acad. xx, 277. *Oreocarya suffruticosa,* Greene.

Infrequent; in canons near the Platte River, July 14, 1894 (No. 477).

Krynitzkia Pattersoni, Gray, Proc. Am. Acad. xx, 278.
    Rare ; in the Laramie Hills, July 7, 1894 (No. 412).

Krynitzkia sericea, Gray, Proc. Am. Acad. xx, 277. *Oreocarya sericea,* Greene.
    Frequent on rocky slopes in the Laramie Hills; June 16, 1894 (No. 255); Uva, July 10, 1894 (No. 389).

Krynitzkia virgata, Proc. Am. Acad. xx, 279.
    Frequent on sandy ridges in the foothills.
    Telephone canon, June 15, 1894 (No. 231); Centennial Valley, June 9, 1895 (No. 1267).

Krynitzkia Watsoni (?) Gray, Proc. Am. Acad. xx, 270.
    These are somewhat doubtfully placed here; collected on wet, shaded rocky ledges in the Centennial Hills, August 17, 1895 (No. 1684).

Mertensia alpina, Don. Syst. iv, 320.
    Strictly alpine from a climatic point of view, but hardly so from that of altitude; very early in the Laramie Hills while freezing nights are still the rule; often in blossom by April 20. (Nos. 33 and 1222).

Mertensia lanceolata, DC. Prodr. x, 88 (1846).
    Very variable as to size and general appearance, but floral characters and the light-green glaucus color constant. It is abundant and frequent in our foothills in two forms : a small form from large, coarse rootstocks, very early, radical leaves few, stem leaves nearly uniform, panicle close and leafy; a much larger form later in the season, usually in copses, stem leaves gradually reduced in size, panicle long and open. 1, Laramie Hills, May 16, 1894 (No. 34); 2, Pole Creek, June 28, 1895 (No. 1234).

Mertensia lanceolata viridis, n. var.
    Root stocks slender, creeping in the crevices among the rocks; radical leaves numerous, long and slender petioled, from oblong to elliptical; cauline leaves oblong, gradually reduced in size; stems few and slender, 5-8 inches high, bearing an open panicle; floral characters those of the species, except that the corolla tube is wider and shorter. *Mertensia lanceolata,* DC.
    This plant is alpine in habitat and may be known by the bright green color of its leaves, which are scarcely scabro-puberulent under a lens.

On rocky ledges near the summit of Laramie Peak, August 7, 1895 (No. 1608).

Mertensia sibirica, Don. Syst. iv, 320.

Quite a large series of specimens from many localities and different altitudes, and of very different general appearance, have all been reduced to this. Environment produces greater differences than is usually conceded. Wet places, along streams and in the mountains; Sybille, July 8, 1894 (No. 408); Garfield Peak, July 29, 1894 (No. 689); Union Pass, August 13, 1894 (No. 1031) and other localities.

Myosotis sylvatica alpestris, Koch.

This beautiful little plant is found in abundance in the alpine regions of our northern mountains. Little Bald Mountain, August 15, 1892 B. C. Buffum; Union Pass, August 11, 1894 (No. 838).

Onosmondium molle, Michx. Fl. Bor. Am. 1 : 133, t. 15 (1803).

Rare and probably confined to the eastern part of the state. Orin Junction, August 1892 ; Platte River, July 14, 1894 (No. 506).

Lithospermum angustifolium, Michx. Fl. Bor. Am. i, 130 (1803).

Very common on the plains and in the mountain valleys. University Campus, June 4, 1894 (No. 174); in its fruiting form July 23, 1895 (No. 1428).

Lithospermum pilosum, Nutt. Journ Acad. Phil. vii, 43 (1834).

Much less frequent than the preceding; Middle Pass, June 20, 1892 ; Gros Ventre River, August 18, 1894 (No. 1090).

CONVOLVULACEÆ.

Ipomea leptophylla, Torr. in Frem. Rep. 95 (1845).

Abundant in the eastern part of the state on the Platte and in the adjacent foothills

Fort Laramie, September 3, 1892 ; Fairbanks, July 14, 1894 (No. 491).

Convolvulus sepium, L. Sp. Pl. 153 (1753).

Collected by B. C. Buffum, August 1, 1891, probably at Cheyenne. Rare in the state.

Evolvulus argenteus, Pursh Fl. Am. Sept. 187 (1814). *E. Nuttallianus,* R. & S.

Infrequent ; dry plains near Uva, July 10, 1894 (No. 398); also at Sheridan, September, 1895.

—12

Cuscuta decora, Engelm. Trans. St. Louis Acad. i. 501 (1859). *C. indecora,* Choisy.

Parasitic on *Alfalfa,* Sheridan Experiment Farm, August 19, 1892.

Cuscuta epilinum, Weihe. Archiv. Apoth. viii. 54 (1824).

On *Alfalfa,* Laramie Experiment Farm, July, 1894 (No. 1210).

SOLANACEÆ.

Solanum rostratum, Dunal. Sol. 234, t. 24 (1813).

Probably abundant on the eastern border.

Fort Laramie, September 5, 1892; Whalen Canon, July 18, 1894 (No. 527). Not yet reported as a bad weed in any locality.

Solanum triflorum, Nutt. Gen. i, 128 (1818).

Particularly annoying as a weed in garden and "truck" patches; small plants, such as *Carrots* and *Parsnips,* must be "weeded" by hand or this weed will completely smother them.

Experiment Farm, September 15, 1894.

Physalis lanceolata, Michx. Fl. Bor. Am. i, 149 (1803).

This genus seems to be confined to the lower altitudes of the north and east; rather plentiful on the Platte and adjoining foothills.

A large form of this species from Blue Grass Hills, August 8, 1894 (No. 365); also a small and more pubescent form from the same place (No. 295).

Physalis lanceolata lævigata, Gray. Proc. Am. Acad. x, 62.

Infrequent; in a canon near the Platte at Fairbanks, July 14, 1894 (No. 478).

SCROPHULARIACEÆ.

Scrophularia Marylandica, L. Sp. Pl. 619 (1753).

Our specimens are far from typical; the thyrsus very narrow, the separate cymes simple; leaves large and truncate at base.

Sybille, July 8, 1894 (No. 317); Pole Creek, June 30, 1895 (No. 1410). Specimens from Garfield Peak, July 29, 1894 (No. 690) have large deltoid-cordate leaves even to the summit of the stem; infloresence only a simple cyme.

Pentstemon acuminatus, Dougl. Lindl. Bot. Reg. t. 1285 (1829).

Abundant in the Laramie range in the foothills, on stony slopes. Laramie Hills, June 7, 1894 (No. 180); Table Mountain, June 28, 1895 (No. 1325).

Pentstemon cæruleus, Nutt. Gen. ii, 52 (1818).

Frequent on the Laramie Plains, especially on sandy ridges bordering on the Laramie River; June 7, 1894 (No. 179); June 18, 1895 (No. 1308).

Pentstemon cæspitosus, Nutt. Gray, Proc. Am. Acad. vi, 66.

A few good specimens of this rather rare plant from Wheatland, June 11, 1892, by B. C. Buffum.

Pentstemon confertus cæruleo-purpureus, Gray, Proc. Am. Acad. vi, 72.

Frequent on subalpine grassy slopes, and much reduced specimens from alpine locations.

Saratoga July 17, 1892; Union Pass, August 11, 1894 (No. 833); Union Peak, August 13, 1894 (No. 1017).

Pentstemon cristatus, Nutt. Gen. ii, 52 (1818).

Common in sandy ravines in the foothills of the Laramie range and probably elsewhere.

Telephone Canon, June 15, 1894 (No. 235); Pole Creek, June 28, 1895 (No. 1346).

Pentstemon glaber, Pursh, Fl. Am. Sept. 738 (1814).

Infrequent and usually only scattering plants.

Sybille, July 8, 1894 (No. 327); Cottonwood Canon, August 4, 1895 (No. 1566).

Pentstemon glaber Utahensis, Watson, Bot. King Surv. 217 (1871).

Slightly variant, but I believe true specimens of this were obtained in three localities: Gros Ventre River, August 18, 1894 (No. 1093); Cummins, July 30, 1895 (No. 1544); Cottonwood Canon, August 4, 1895 (No. 1579).

Pentstemon glaucus, Graham, Edinb. Phil. Journ. 1829, 348.

Herbarium specimens fail to do justice to this singularly beautiful plant, as shape and color are both largely lost. It occurs in the alpine region of all our ranges.

Tetons, August 24, 1894 (No. 1001); Laramie Peak, August 7, 1895 (No. 1619); La Plata Mines, August 22, 1895 (No. 1792).

Pentstemon humilis, Nutt.
    Exceedingly abundant in the Laramie range at the foot of rocky ledges and in stony ravines.
    Pole Creek, June 2, 1894 (No. 131); Table Mountain, June 27, 1895 (No. 1322).

Pentstemon laricifolius, Hook. & Arn. Bot. Beechy. 376.
    Abundant on stony ridges on the Laramie Plains and their foothills. July 7, 1894 (No. 419); July 27, 1895 (No. 1442).

Pentstemon secundiflorus, Benth. DC. Prodr. x, 324.
    Infrequent; specimens by Mr. Hartley from near Sherman, July 9, 1891.

Pentstemon strictus, Benth. DC. Prodr. x, 324.
    In stony ravines in the hills; Sybille, July 8, 1894 (No. 402); Cummins, July 28, 1895 (No. 1472).

Collinsia parviflora, Lindl. Bot. Reg. t. 1082 (1827).
    Very abundant in sandy loam soil on creek banks.
    Pole Creek, June 2, 1894 (No. 46); Centennial Valley, July 17, 1895 (No. 1685).

Mimulus alsinoides, Benth. DC. Prodr. x, 351.
    Centennial Hills, August 17, 1895 (No. 1683).

Mimulus floribundus, Lindl. Bot. Reg. xiv, t. 1125 (1828).
    Infrequent: good, but diminutive; from a wet ravine near Cummins, July 28, 1895 (No. 1515).

Mimulus glabratus Jamesii, Gray, Syn. Fl. Suppl. 447.
    Infrequent; Whalen Canon, July 19, 1894 (No. 543).

Mimulus Langsdorfii Tilingi, Greene, Journ. Bot. for Jan., 1895.
    Our commonest Mimulus; springy places in the hills and mountains.
    Garfield Peak, July 29, 1894 (No. 688); Centennial Valley, August 16, 1895 (No. 1670).

Mimulus Lewisii, Pursh, Fl. ii. 427 (1814).
    On stony, wet ground near creeks; infrequent.
    Teton Mountains, August 21, 1894 (No. 942); Centennial Valley, August 16, 1895 (No. 1672).

Mimulus rubellus, Gray, Bot. Mex. Bound. 116.
    Rare, or at least rarely observed; our plants small and inconspicuous.

On a naked, gravelly hillside, Centennial, June 9, 1895 (No. 1287).

Synthyris plantaginea, Benth. DC. Prodr. x, 455 (1846).
Some specimens from the Teton Mountains are doubtfully placed here. August 21, 1894 (No. 986).

Synthyris rubra, Benth. l. c. *Wulfenia rubra,* (Hook.) Greene.
Of very frequent occurrence in sage brush valleys ; variable as to foliage.
Telephone Canon, May 12, 1894 (No. 29); Laramie Hills, June 5, 1895, (No. 1242).

Veronica alpina, L. Sp. Pl. 11 (1753).
Frequent in subalpine stations ; wet, grassy slopes.
Bald Mountain, August 15, 1892 ; Union Pass, August 11, 1894 (No. 831); Centennial Hills, August 19, 1895 (No. 1740).

Veronica Americana, Schwein. Benth. in DC. Prodr. x, 468 (1846).
In all streams and springs.
Sybille Creek, July 8, 1894 (No. 403) ; noted in many other localities.

Veronica peregrina, L. Sp. Pl. 14 (1753).
In boggy places ; Lander, August 3, 1894 (No. 698); Centennial Valley, August 25, 1895 (No. 1854).

Veronica serpyllifolia, L. Sp. Pl. 12 (1753).
On the grassy banks of our little mountain brooks ; frequent.
Horse Creek, June 9, 1894 (No. 192).

Castilleia flava, Watson, Bot. King Surv. 230 (1871).
Of rather frequent occurrence at 7,000-8,000 ft. in the hills.
Laramie Hills, July 7, 1894 (No. 352); Table Mountain, June 27, 1895 (No. 1338).

Castilleia linariæfolia, Benth. DC. Prodr. x, 520.
Hardly subalpine ; frequent on sandy, grassy slopes.
Laramie Hills, July 7, 1894 (No. 352); Laramie Peak, August 6, 1895 (No. 1570).

Castilleia miniata, Dougl. Hook. Fl. ii, 106.
I have placed here a number of specimens, some of which rather doubtfully. Most of them were secured on wet stream banks and I think the following numbers, at least, are right. Lander, August 4, 1894 (No. 712); Union Pass, August 11,1894 (No. 835).

Castilleia minor, Gray, Bot. Cal. i, 573 (1876).

Rare; observed but once. Fort Washakie, August 5, 1894 (No. 744).

Castilleia pallida, Kunth. Syn. Pl. Equin. ii, 100.

On the shaded banks of mountain streams, at 8,000 ft. and upward.

Cummins, July 28, 1895 (No. 1461); observed at a number of other localities.

Castilleia pallida occidentalis, Gray, Bot. Cal. i, 573 (1876).

Some specimens from Union Peak I am unable to place elsewhere; August 13, 1894 (No. 1011).

Castilleia pallida septentrionalis, Gray, l. c.

Undoubted specimens of this were observed in a number of alpine and subalpine stations. This prefers the wet banks of wooded streams or the shores of alpine lakes.

Centennial Hills, August 18, 1895 (No. 1726); La Plata Mines, August 23, 1895 (No. 1808).

Castilleia parviflora, Bong. Veg. Sitch. 158 (1831).

Most frequent and earliest; in sandy loam soil among the sage brush.

Pole Creek, June 2, 1894 (No. 120); Centennial Valley, June 9, 1895 (No. 1291).

Orthocarpus luteus, Nutt. Gen. ii, 56 (1818).

Frequent in wet, sandy soil, especially along streams.

Whalen Canon, July 16, 1894 (No. 534); Lander, August 3, 1894 (No. 744); Cummins, July 30, 1895 (No. 1537).

Orthocarpus pallescens, Gray, Am. Journ. Sci. ser. 2, xxxiv, 339.

From type locality, probably not far from the place where Parry collected it.

Gros Ventre River, August 15, 1894 (No. 900).

Orthocarpus pilosus, Watson, Bot. King Surv. 234 (1871).

This rare plant, determined for me by Dr. Rose, was collected by B. C. Buffum, on Three Mile Creek, June 26, 1892.

Cordylanthus ramosus, Nutt. Gen. ii, 57 (?).

Very abundant on the alkali-clay hills adjacent to the Wind River; at Dubois, August 10, 1894 (No. 711).

Pedicularis bracteosa, Benth. Hook. Fl. ii, 110 (1838).
   Probably confined to our northwest areas.
   Union Pass, August 11, 1894 (No. 834).
Pedicularis crenulata, Benth. DC., Prodr. x, 568.
   Abundant; growing in clumps in high meadow lands.
   Laramie, June 19, 1892; Cummins, July 30, 1895 (No. 1528).
Pedicularis Grœnlandica, Retz. Fl. Scand. Ed. 2, 145 (1795).
   Frequent and abundant on subalpine mountain streams.
   Warm Spring Creek, August 19, 1894 (No. 806); Centennial Hills, August 19, 1895 (No. 1739).
Pedicularis Parryi, Am. Journ. Sci. ser. 2, xxxiii, 250.
   Probably rare; a few specimens only, from Union Peak, August 13, 1894 (No. 1033).
Pedicularis procera, Gray, Am. Journ. Sci. ser. 2, xxxiv, 251.
   Rare; in a wooded canon leading to the Laramie River, at Cummins, July 30, 1895 (No. 1550).
Pedicularis racemosa, Dougl. Hook. Fl. ii, 108 (1838).
   This comes as near as any to marking a zonal belt; encountered at about 9,000 ft. in all our mountain ranges visited, and found up to timber line.
   Bald Mountain, August 17, 1892; Union Pass, August 11, 1894 (No. 830); noted at Laramie Peak and in the Medicine Bow Mountains.

OROBANCHACEÆ.

Aphyllon fasciculatum, Gray, Syn. Fl. 2; Part 1, 312 (1878). *Thalesia fasciculata*, Britton.
   Frequent and occasionally abundant.
   Pine Creek, July 18, 1892; Pole Creek, July 1, 1895 (No. 1360), hosts,—*Artemisia frigida* and *A. Canadensis*.
Aphyllon uniflorum, T. & G. Gray, Man. 290 (1848).
   Rare; Gros Ventre River, August 22, 1894 (No. 1071), host not noted.

VERBENACEÆ.

Lippia cuneifolia, Steud. Torr., in Marcy's Rep. 293, t. 17 (1853).
   Laramie River bottom lands near Uva; not observed elsewhere. July 10, 1894 (No. 387).

Verbena bracteosa, Michx. Fl. Bor. Am. ii, 13 (1803).

   This is a weedy plant, thriving equally well in cultivated and uncultivated grounds.

   Blue Grass Hills, July 8, 1894 (No. 320); Laramie Peak, August 8, 1895 (No. 1652).

Verbena stricta, Vent. Hort. Cels, t. 53 (1800).

   Probably frequent in the lower altitudes of the northeast; not observed except in the Platte Valley, near Fairbanks, July 14, 1894 (No. 505). A form with several large spikes was obtained in Whalen Canon, July 18, 1894 (No. 538).

LABIATÆ.

Mentha Canadensis, L. Sp. Pl. 576 (1753).

   Frequent on all water courses and about springs.

   Whalen Canon, July 18, 1894 (No. 546); specimens from a large number of other localities.

Lycopus lucidus, Turcz. Benth. in DC. Prodr. xii, 178 (1848).

   Infrequent; Popo Agie River, August 2, 1894 (No. 735).

Lycopus sinuatus, Ell. Bot. S. C. and Ga. i, 126 (1816).

   Wet grounds about springs and ponds; frequent.

   Laramie, July, 1891; Whalen Canon, July 19, 1894 (No. 542).

Hedeoma Drummondii, Benth. Lab. Gen. and Sp. 368 (1834).

   Occasional in dry loam soil.

   Whalen Canon, July 19, 1894 (No. 548).

Hedeoma Reverchoni, Gray, Syn. Fl. 363.

   Rare; its collection in this state extends the range of this plant northward very much; good specimens from Pole Creek, June 30, 1895 (No. 1374).

Salvia lanceolata, Willd. Enum. 37 (1809).

   Only occasionally on rather dry hillsides and plains.

   Blue Grass Hills, July 9, 1894 (No. 374); Centennial Valley, August 20, 1895 (No. 1395).

Monarda fistulosa, L. S. Pl. 22 (1753).

   Probably confined to the eastern portion of the state; thus far noted only in the region east of Laramie Peak.

   Fairbanks, July 14, 1894 (No. 508); Cottonwood Canon, August 4, 1895 (No. 1577).

Lophanthus urticifolius, Benth. Bot. Reg. xv, sub. t. 1282 (1829).
Vleckia urticifolia, (Benth.) Holzinger.
> Very rare; possibly only in the northwest.
> Snake River, August 22, 1894 (No. 984).

Dracocephalum parviflorum, Nutt. Gen. ii, 35 (1818).
> Principally in abandoned fields.
> Sybille Creek, July 8, 1894 (No. 326); Laramie Peak, August 8, 1895 (No. 1648).

Scutellaria galericulata, L. Sp. Pl. ii, 599 (1753).
> Not infrequent in wet places.
> Fairbanks, July 13, 1894 (No. 470); Centennial Valley, August 20, 1895 (No. 1760).

Scutellaria resinosa, Torr. Ann. Lyc. N. Y. ii, 232 (1827).
> On abrupt banks along streams, but in comparatively dry ground.
> Pole Creek, June 2, 1894 (No. 94); June 29, 1895 (No. 1365).

Brunella vulgaris, L. Sp. Pl. 600 (1753).
> Very infrequent in the parts of the state collected
> Specimens from Wolf Creek, August 18, 1892, B. C. Buffum.

Physostegia parviflora, Nutt. Benth. in DC. Prodr. 12, 434 (1848).
> On the borders of lakes and ponds, growing even in the edge of the water.
> Bull Lake, August 8, 1894 (No. 732); Laramie River, near Ione Ranch, August 10, 1895 (No. 1666).

Stachys palustris, L. Sp. Pl. 580 (1753).
> Frequent on sandy creek banks which are covered with undershrubs and weedy plants.
> Wind River, August 8, 1894 (No. 857); Cummins, July 28, 1895 (No. 1484); noted in many other places.

PLANTAGINACEÆ.

Plantago eriopoda, Torr. Ann. Lyc. N. Y. ii, 237 (1827).
> Common on wet alkali flats about Laramie and in similiar locations elsewhere; June 1, 1894 (No. 42). *Plantain.*

Plantago lanceolata, L. Sp. Pl. 113 (1753).
> A recently introduced weed on the Lander Experiment Farm; September, 1895, by J. S. Meyer.

Plantago major, L. Sp. Pl. 112 (1753).

Received from Sheridan as one of the weeds upon the Experiment Farm, September, 1895, by J. F. Lewis.

Some specimens from Wolf Creek, apparently native there, collected August 8, 1892, by B. C. Buffum, are doubtfully placed here. Spike very slender and leaves almost acute.

Plantago Patagonica gnaphalioides, Gray, Man. Ed. 2, 269 (1856). *Plantago Purshii,* R. & S.

Frequent on dry, gravelly hillsides in the Laramie range. Observed only in the eastern part of the state.

Laramie, August, 1891; Wheatland, July, 1891; Table Mountain, June 28, 1895 (No. 1356).

Plantago Tweedyi, Gray, Syn. Fl. II, Part i, 390 (1886).

Three specimens only of this very rare plant were secured on a grassy hillside near the La Plata Mines, August 21, 1895 (No. 1798).

I am not aware that any other specimens have been collected since it was originally collected on the Yellowstone River by Mr. Frank Tweedy.

Oxybaphus angustifolius, Sweet. Hort. Brit. 429 (1830). *Allionia linearis,* Pursh.

Never abundant; only scattering specimens, often in cultivated ground.

Wheatland, July 9, 1894 (No. 379); Cottonwood Canon, August 4, 1895 (No. 1560); also from Sheridan, September, 1895.

Oxybaphus hirsutus, Choicy in DC. Prodr. xiii, part 2, 433 (1849). *Allionia hirsuta,* Pursh.

Rare; noted but once; Whalen Canon, July 18, 1894 (No. 515).

Oxybaphus nyctagineus, Sweet. Hort. Brit. 429 (1830). *Allionia nyctaginea,* Michx.

Very rare, unless it be in the northeastern part of the state. Fairbanks, July 14, 1894 (No. 469).

Abronia fragrans, Nutt. Hook. Kew. Journ. Bot. v, 261 (1853).

Frequent on the sandy plains of the eastern part of the state.

Inyan Kara Divide, August 29, 1892; Platte River, July 14, 1894 (No. 464); Cummins, July 28, 1895 (No. 1473).

Abronia micrantha, Torr.

Very rare; on the plains and hillsides; by children called the *Sand Flower.*

Near Willow Creek, July 22, 1894 (No. 630).

ILLECEBRACEÆ.

Paronychia Jamesii, T. & G. Fl. N. A. i, 170 (1838).

Frequent on dry, open slopes in the Laramie range.

Fairbanks, July 14, 1894 (No. 451); also near Laramie Peak.

Paronychia pulvinata, Gray, Proc. Acad. Phila. 1863, 58.

Infrequent and strictly alpine; on the naked summits of the Medicine Bow Mountains, August 23, 1895 (No. 1824).

Paronychia sessiliflora, Nutt. Gen. i, 160 (1818).

Common on the slopes about Laramie Peak, August 8, 1895 (No. 1638); observed in some other localities.

Paronychia sp.

Numbers 349, 461, 1331 and 1656 have for the present been laid aside. They represent at least two species not given above, but for want of sufficient literature they are now passed over.

AMARANTACEÆ.

Amarantus albus, L. Sp. Pl. Ed. 2, 1404 (1763).

A troublesome weed in most of our fields and gardens.

Popo Agie River, August 3, 1894 (No. 740); Sheridan Experiment Farm, September 1895.

Amarantus blitoides, Wats. Proc. Am. Acad. xii, 273 (877).

This too would find a place even in a very short list of the worst weeds

Laramie, September 24, 1894 (No. 1176); Sheridan, September, 1895.

Amarantus chlorostachys, Willd. *A. hybridus,* L.

Some specimens, apparently this, were collected by B. C. Buffum, at Inyan Kara Divide, August 30, 1892.

Amarantus retroflexus, L. Sp. Pl. 991 (1753).

Where weeds run riot this will always be found.

Laramie, October, 1893; Lander, August 3, 1894 (No. 715)

Amarantus Torreyi, Benth. Wats. Bot. Cal. ii, 42 (1880).

Infrequent; near Rawlins, June 29, 1892.

CHENOPODIACEÆ.

Cycloloma platyphyllum, Moq. Enum. Chenop. 18 (1840). *C. atriplicifolium*, Coulter.

>Frequent in the sandy canons leading to the Platte, near Fairbanks, July 14, 1894 (No. 471).

Monolepsis chenopodioides, Moq. DC. Prodr. xiii, part 2, 85 (1849). *Monolepsis Nuttalliana*, Greene.

>Frequent in saline ground.
>Laramie, August, 1893; Boulder Creek, August 26, 1894 (No. 1103).

Chenopodium album, L. Sp. Pl. 219 (1753).

>In waste and cultivated ground.
>Laramie, October, 1893; Sheridan Experiment Farm, September, 1895. *Pigweed.*

Chenopodium capitatum, Watson, Bot. Cal. ii, 48 (1880).

>Infrequent; leaves only, from Cummins, July 27, 1895 (No. 1448).

Chenopodium Fremontii, Watson, Bot. King's Exp. 287 (1871).

>Abundant on dry hillsides and on rocky shelving ledges.
>Fairbanks, July 13, 1894 (No. 440).

Chenopodium glaucum, L. Sp. Pl. 220 (1753).

>Apparently indigenous in places, but scattering everywhere as a weed.
>Poison Spider Creek (native?), July 27, 1894 (No. 625); among the weeds at Laramie and Sheridan.

Chenopodium leptophyllum, Nutt. Moq. DC. Prodr. xiii, part 2, 71 (1849).

>Not infrequent near the Platte and its tributaries; Laramie, August 15, 1893; Platte River, July 14, 1894 (No. 483).

Chenopodium leptophyllum subglabrum, Wats. Proc. Am. Acad. ix, 95.

>Rare; Willow Creek, July 22, 1894 (No. 628).

Chenopodium olidum, Watson, Proc. Am. Acad. ix, 96.

>In saline ground on Poison Spider Creek, July 27, 1894 (No. 622); also sent from the Sheridan Experiment Farm as a weed, 1895.

Chenopodium rubrum humile, Watson, Bot. Cal. ii, 48.

>Some specimens from Laramie, by B. C. Buffum, June 23, 1894, seem to belong here. Not observed elsewhere.

Atriplex argentea, Nutt. Gen. i, 198 (1818).
Throughout the state; reported as a weed from some localities.
Big Muddy Creek, July 24, 1894 (No. 640); Meadow Creek, August 9, 1894 (No. 791); Sheridan, September, 1895.

Atriplex canescens, James, Wats. Proc. Am. Acad. ix, 110 (1874).
Not frequent; observed only on the east side of the Laramie range.
Wheatland, August 11, 1891; Platte Hills, Fairbanks, July 14, 1894 (No. 463).

Atriplex confertifolia, Watson, l. c.
A common shrub on the alkali deserts of the central and southwestern parts of the state. Frequently called *White Sage* and is said to be freely eaten by antelope and sheep during the winter months.
Rock Springs, September 9, 1893; Bessemer, July 26, 1894 (No. 613).

Atriplex expansa, Watson, Proc. Am. Acad. ix, 116 (1874).
Very abundant in the vicinity of alkali lakes.
Howell Lakes, October 1, 1894 (No. 1164).

Atriplex Nuttallii, Watson, l. c.
Luxuriant in strongly saline ground.
Alkali Desest, July 2, 1891; Howell Lakes, October 1, 1894 (No. 1167); alkali flats near Sweetwater River, September 10, 1894 (No. 1192).

Atriplex patula hastata, Gray, Man. Ed. 5, 409 (1867). *A. hastata,* L.
Abundant near Laramie and probably in many other localities. September 1, 1895 (No. 1866). The form with ovate entire leaves was secured at the same time.

Atriplex truncata, Gray, Proc. Am. Acad. viii, 398.
I think there is little doubt of the correctness of this reference.
Rather immature specimens from Poison Spider Creek, July 26, 1894 (No. 621); older at Howell Lakes, October 1, 1894 (No. 1166).

Eurotia lanata, Moq. Enum. Chenop. 81 (1840).
Very frequent and abundant in many localities.
Fairbanks, July 14, 1894 (No. 488); Laramie, September 12, 1894 (No. 1134).

Salicornia herbacea, L. Sp. Pl. 3 (1762).

Abundant on the shores of alkali lakes, even on the encrusted banks where nothing else will grow.

Howell Lakes, October 1, 1894 (No. 1162); Laramie, September 3, 1895 (No. 1869).

Sueda depressa, Watson, Bot. King's Exp. 294 (1871).

Not noted except at Howell Lakes, October 1, 1894 (No. 1163).

Sueda depressa erecta, Watson, Proc. Am Acad. ix, 90 (1874).

Not frequent; on an alkali bog on Poison Spider Creek, July 27, 1894 (No. 623).

Sueda diffusa, Watson, Proc. Am. Acad. ix, 88 (1874).

This is common on wet, saline ground.

Popo Agie River, August 4, 1894, (No. 717); Sheridan, September, 1895, J. F. Lewis.

Sueda Torreyana, Watson, Proc. Am. Acad. ix, 88 (1874).

Only collected at Howell Lakes, but probably elsewhere as well. October 1, 1894 (No. 1168).

Salsola Kali Tragus, Moq. in DC. Prodr. xiii, part 2, 187 (1849).

This, the much talked of *Russian Thistle*, is already quite widely distributed in the northeastern part of the state. Other plants are frequently mistaken for it and so it has often been reported from localities in which it does not exist. There is no evidence to show that it is found on the line of the Union Pacific railroad except at Cheyenne. On the lines of the Burlington & Misssouri and the Elkhorn roads, however, I am reliably informed that it is widely distributed.

Specimens from Cheyenne, October, 1894; from Lusk and Fredericks in 1895.

Sarcobatus vermiculatus, Torr., Emory's Rep. 150 (1848).

The most characteristic shrub of the saline plains and foothills. Sweetwater River, September 9, 1894 (No. 1182); Howell Lakes, October 1, 1894 (No. 1161). *Grease Wood*.

POLYGONACEÆ.

Eriogonum alatum, Torr. Sitgreaves Rep. t. 8 (1853).

Noted only on the foothills of the Laramie range and the adjoining plains; abundant.

Cheyenne, August 11, 1891; Laramie Hills, July 9, 1894 (No. 416).

Eriogonum annuum, Nutt. Trans. Am. Phil. Soc. (II) 5: 164 (1833-37).
Infrequent in parts of state collected.
Inyan Kara Divide, September 1, 1892; Willow Creek, July 20, 1894 (No. 566).

Eriogonum brevicaule, Nutt.
Immensely abundant throughout the state on clayey, gravelly ridges and plains.
Laramie Hills, July 8, 1894 (No. 303); Wind River, August 8, 1894 (No. 722); from several other localities.

Eriogonum cæspitosum, Nutt. Journ. Acad. Phil. vii, 50 (1834).
Very rare; Union Pass, August 14, 1894 (No. 890).

Eriogonum cernuum, Nutt. Journ. Acad. Phil. (II) 1: 162 (1848).
Widely distributed and abundant.
Laramie Hills, July 7, 1894 (No. 363); Wind River, August 8, 1894 (No. 726).

Eriogonum chrysocephalum, Gray, Proc. Am. Acad. xi, 101.
Rare; possibly confined to the Pacific Slope.
Bacon Creek, August 15, 1894 (No. 903).

Eriogonum flavum, Nutt., Fras. Cat. (1813).
Common on hillsides in the Laramie and Medicine Bow ranges.
Laramie Hills, July 8, 1894 (No. 415); Cummins, July 27, 1895 (No. 1443); a very large form with compound umbels from Laramie Peak, August 5, 1895 (No. 1572).

Eriogonum flavum crassifolium, Benth.
Frequent in the Laramie foothills. Name communicated by Dr. Robinson; citation not at hand.
Laramie Hills June 6, 1894 (No. 186).

Eriogonum heracleoides, Nutt. Journ. Acad. Phila. vii, 49 (1834).
Infrequent; our specimens are not typical but they undoubtedly must be placed here.
Wallace Creek, July 29, 1894 (No. 677).

Eriogonum microthecum, Nutt. Journ. Acad. Phila. (II) 1: 162 (1848).
Both frequent and abundant.
Laramie at various times (Nos. 329 and 1138); Sweetwater, September 12, 1894 (No. 1188).

Eriogonum ovalifolium, Nutt. Journ. Acad. Phila. vii, 50 (1834).
   In dry sandy soil on the plains; frequent.
   Laramie, May 29, 1894 (No. 70); Hutton ranch, June 19, 1894 (No. 278).

Eriogonum subalpinum, Greene, Pitt. xiii, part 13, 18 (1896).
   This recently described species, founded, seemingly, on unimportant characters I believe to be a valid one. The eight specimens in our herbarium labelled *E. umbellatum* fall readily into the two groups made by Prof. Greene. At different times the writer has brought in the two forms from the field fully convinced that they were different and yet unable to separate them by any marked botanical characters, though so readily distinguished in the field.
   Laramie Hills, July 7, 1894 (No. 346); Tetons, August 22, 1894 (No. 975); others from Laramie.

Eriogonum umbellatum, Torr. Ann. Lyc. N. Y. ii, 241 (1828).
   Pole Creek, July 2, 1895 (No. 1419); Laramie Hills at various times.

Polygonum amphibium, L. Sp. Pl. 361 (1753).
   Observed only in the Platte River near Fairbanks, but undoubtedly found elsewhere. July 15, 1894 (No. 551).

Polygonum aviculare, L. Sp. Pl. 362 (1753).
   A common dooryard weed.
   Laramie, September 24, 1894 (No. 1173).   *Goose Grass*.

Polygonum bistortoides, Pursh, Fl. i, 217 (1814).
   Frequent in grassy valleys.
   Bald Mountain, August 15, 1892; Union Pass, August 11, 1894 (No. 1024).

Polygonum convolvulus, L. Sp. Pl. 364 (1753).
   Waste and cultivated ground; Experiment Farm, September 15, 1894 (No. 1140); Popo Agie river, August 1, 1894 (No. 714).

Polygonum Douglasii, Greene, Bull. Cal. Acad. iii, 125 (1885).
   In dry ravines; Sybille Creek, July 7, 1894 (No. 307); Laramie Peak, August 7, 1895 (No. 1591).

Polygonum erectum, L. Sp. Pl. 363 (1753).
   Infrequent; on the bank of Blue Grass Creek, July 8, 1894 (No. 364).

Polygonum hydropiper, ~~hydropiper~~ [*lapathifolium Luсanum*], L. Sp. Pl. 361 (1753).
 In a wet place on the banks of a small stream near Lusk, July 21, 1894 (No. 576).

Polygonum minimum, Watson, Bot. King Surv. v, 315 (1871).
 I refer our specimens to this with some doubt; they are larger than the type and less pubescent.
 La Plata Mines, August 24, 1895 (No. 1833).

Polygonum nodosum, Pers. Syn. i, 440 (1805). *P. lapathifolium nodosum,* (Pers.) Small.
 From Wheatland, August 1892, by B. C. Buffum.

Polygonum Persicaria, L. Sp. Pl. 362 (1753).
 As a weed on the Sheridan Experiment Farm, September 1895, by J. F. Lewis. *Smart Weed.*

Polygonum ramosissimum, Michx. Fl. Bor. Am. i, 237 (1803).
 Frequent and variable, some of the specimens approaching *P. Tenue.*
 Sugg's Road, August 20, 1892; Wallace Creek, July 29, 1894 (No. 671); Green River, August 14, 1894 (No. 888).

Polygonum viviparum, L. Sp. Pl. 360 (1753).
 Frequent in the higher mountains.
 Fort Laramie, September 5, 1892; Union Pass, August 11, 1894 (No. 842).

Oxyria digyna, Campd. Rum. 155, t. 3, f. 3 (1819)
 Very frequent in shaded places and under overhanging cliffs, 8,000-9000 ft.
 Garfield Peak, July 29, 1894 (No. 662); Laramie Peak, August 7, 1895 (No. 1620).

Rumex Crispus, L. Sp. Pl. 335 (1753).
 Platte River, July 14, 1894 (No. 495); Laramie, September 20, 1894 (No. 1147).

Rumex maritimus, L. Sp. Pl. 335 (1753).
 In saline soil; not frequent.
 Clark's Ranch on Wind River, August 10, 1894 (No. 769); Sheridan, September 1895, J. F. Lewis.

Rumex occidentalis, Watson, Proc. Am. Acad. xii, 253 (1876).
 Under this name are a number of specimens, some immature, and only the following typical:

—13

Wallace Creek, July 29, 1894 (No. 561); Centennial Valley, August 16, 1895 (No. 1752).

Rumex paucifolius, Nutt. Ms. in Herb. Gray. *R. Geyeri*, (Meisn.) Trelease.

Abundant in subalpine parks ; Union Pass, August 11, 1894 (No. 855); also observed in the Medicine Bow Mountains.

Rumex salicifolius, Weinm. Flora, iv, 28 (1813).

Frequent and somewhat variable in general appearance.

Whalen Canon, July 17 1894 (No. 561); Laramie, July 23, 1895 (No. 1449).

Rumex venosus, Pursh, Fl. Am. Sept. 733 (1814).

Frequent about sand dunes on the plains.

Laramie, June 3, 1894 (No. 156).

ELÆAGNACEÆ.

Eleagnus argentea, Pursh, Fl. Am. Sept. 114 (1814).

Very frequent on the Wind Rivers.

Fort Washakie, August 5, 1894 (No. 703).

Shepherdia argentea, Nutt. Gen. ii, 240 (1818). *Lepargyræa argentea*, (Nutt.) Greene. *Buffalo Berry*.

Common on the Platte ; Bessemer, July 26, 1894 (No. 636).

Shepherdia Canadensis, Nutt. Gen. ii, 241 (1818). *Lepargyræa Canadensis*, (L.) Greene.

Common on moist and partially shaded slopes in the mountains. Laramie Hills, May 12, 1894 (No. 25). Garfield Peak, July 29, 1894 (No. 687); Laramie Peak, August 5, 1895 (No. 1585).

LORANTHACEÆ.

Arceuthobium Americanum, Engelm. Pl. Lindl. ii, 214 (1850).

Razoumofskya Americana, (Engelm.) Kuntze.

Infrequent ; at Keystone, Medicine Bow Mountains, 1893, W. C. Knight.

SANTALACEÆ.

Comandra pallida, DC. Prodr. xiv, 636 (1857).

Frequent on dry slopes throughout the state.

Table Mountain, June 2, 1894 (No. 105); Gros Ventre River, August 18, 1894 (No. 1091).

EUPHORBIACEÆ.

Euphorbia dentata, Michx. Fl. Bor. Am. ii, 211 (1803).
    Infrequent; observed but once; Hartville, July 15, 1894 (No. 549).

Euphorbia dictyosperma, Fisch. and Mey. Ind. Sem. Hort. Petrop. ii, 37 (1835).
    On sand dunes near Mexican Mines, July 21, 1894 (No. 581).

Euphorbia Fendleri, T. & G. Pac. R. R. Rep. ii, 175 (1855).
    Our *Euphorbias* seem to be confined to the warm sandy plains and canons of the eastern part of the state, particularly to the region of the Platte. This from Fairbanks, July 14, 1894 (No. 472).

Euphorbia glyptosperma, Engelm. Bot. Mex. Bound. Serv. 187 (1859).
    Quite frequent; Wheatland, June 29, 1894; Blue Grass Hills, July 8, 1894 (No. 370).

Euphorbia hexagona, Nutt. Spreng. Syst. iii, 791 (1826).
    In a sandy canon leading to the Platte, Fairbanks, July 13, 1894 (No. 436).

Euphorbia montana, Engelm. Bot. Mex. Bound. Surv. 192 (1859).
    This is probably common throughout the state on stony, gravelly hillsides.
    Inyan Kara Divide, August 29, 1892; Whalen Canon, July 18, 1894 (No. 529); Pole Creek, July 1, 1895 (No. 1400).

Euphorbia petaloidea, Engelm. l. c.
    Not infrequent on sandy river bottoms.
    Uva, July 10, 1894 (No. 399); Willow Creek, July 22, 1894 (No. 631).

Croton Texensis, Muell. Arg. in DC. Prodr. xv, part 2, 692 (1862).
    On sandy river bottoms in the eastern part of the state.
    Fairbanks, July 12, 1894 (No. 428); also at Fort Laramie.

URTICACEÆ.

Humulus Lupulus, L. Sp. Pl. 1028 (1753).
    Not infrequent in copses on river bottoms and in ravines.
    Whalen Canon, July 18, 1894 (No. 513). *Wild Hops.*

Urtica Breweri, Watson, Proc. Am. Acad. x, 348 (1875).
    Abundant in Centennial Valley; not observed elsewhere; June 8, 1895 (No. 1273); August 25, 1895 (No. 1862). *Nettle.*

Urtica gracilis, Ait. Hort. Kew. iii, 341 (1789).
   Exceedingly abundant in copses on most streams.
   Mexican Mines, July 20, 1894 (No. 590).
Parietaria Pennsylvanica, Muhl. Willd Sp. Pl. iv, 955 (1805).
   Infrequent; Fairbanks, July 13, 1894 (No. 441).

CUPULIFERÆ.

Betula glandulosa, Mich. Fl. Bor. Am. ii, 180 (1803).
   On the stony banks of subalpine streams.
   Warm Spring Creek, August 10, 1894 (No. 799). *Mountain Birch.*
Betula occidentalis, Hook. Fl. Bor. Am. ii, 155 (1839).
   Supposedly common on our streams, especially in the northeast, but only collected at Laramie Peak, August 8, 1895 (No. 1647). *Western Birch.*
Alnus incana virescens, Wats. Bot. Cal. ii, 81 (1880).
   Frequent and abundant on our mountain streams, sometimes attaining the size of small trees.
   Big Wind River, August 8, 1894 (No. 728); Cummins, July 29, 1895 (No. 1531); also in Centennial Valley. *Alder.*
Quercus undulata, Torr. Ann. Lyc. N. Y. ii, 248 (828).
   This is reported abundant in the Black Hills, to which region it is probably confined; so far as known it is our only *Oak.*
   Sundance, July 22, 1891, by B. C. Buffum.

SALICACEÆ.*

Salix arctica petræa, Anders.
   Infrequent; not observed by the writer.
   Collected by B. C. Buffum, Bald Mountain, August 15, 1892.
Salix alba x fragllis, Wiemner.
   This Hybrid is among those in the City Park, October 1, 1894 (No. 1199).
Salix amygdaloides, Anders. Ofv. Handl. Vet. Akad. 1858, 114 (1858).
   Typical specimens from Snake River, August 21, 1894 (No. 978). My specimens from Big Popo Agie River, August 2, 1894 (No. 738)

---

* The *Willows* were in large part determined by the late Mr. Bebb; those collected in 1895 only, by Mr. M. L. Fernald.

were designated by Mr. Bebb as "*forma sat singularis*," on account of the occurrence in the same individual of the following peculiarities: Leaves unusually short and broad, coarsely serrate and very conspicuously stipulate.

Salix candida, Fleugge, Willd. Sp. Pl. iv, 708 (1806).
Observed only in Centennial Valley, August 20, 1895 (No. 1755).

Salix chlorophylla, Anders. Monogr. 138 (1867).
Widely distributed; Union Pass, August 13, 1894 (No. 1029); Centennial Valley, August 19, 1895 (No. 1741).

Salix cordata, Muhl. Neue Schr. Ges. Nat. Fr. Berlin, iv, 236 (1803).
Collected but once; Laramie Peak, August 6, 1895 (No. 1598).

Salix curtiflora, (Anders.) Bebb.
This very rare plant from two localities; not reported from this state before, I think, except by Mr. Tweedy.
Boulder Creek, August 26, 1894 (No. 1122); Centennial Valley, August 19, 1895 (No. 1754).

Salix desertorum, Rich. App. Frank. Journ. 371 (823).
On a hummocky heath in Centennial Valley, August 20, 1895 (No. 1759).

Salix desertorum elata, Anders.
Rare, alpine; Union Peak, August 13, 1894 (No. 1072); LaPlata Mines, August 22, 1895 (No. 1818).

Salix desertorum fruticulosa, Anders.
Infrequent in an open, heathlike park in Union Pass, August 11, 1894 (No. 943).

Salix flavescens, Nutt. Sylv. i, 65 (1842-53).
This beautiful species is the first to put out its blossoms of all our willows; the leaves follow very tardily.
Thus far noted only in the Laramie range; Pole Creek, May, 12, 1894 (No. 24) and at various other times; Laramie Peak, August 6, 1895 (No. 1586).

Salix lasiandra Fendleriana, (Anders.) Bebb.
Not infrequent; Popo Agie River, August 2, 1894 (No. 737); Pole Creek, June 30, 1895 (No. 1434).

Salix leucosericea, Nob. n. sp.
In communicating this name to me Mr. Bebb made the following comments upon the species: " It will shortly appear as above in a

government report. This is the Rocky Mountain or plateau member of a group which has for its eastern or Atlantic coast representatives, *S. sericea* and *S. petiolaris*, and for the Pacific coast *S. macrocarpa.*"

Rare; observed only on Boulder Creek, August 26, 1894 (No. 1123).

Salix longifolia, Muhl. Neue Schr. Ges. Nat. Fr. Berlin, iv, 238 (1803).
On sand bars and creek banks everywhere, common and variable.
Laramie, June 16, 1894 (No. 245); near Lander, August 3, 1894 (No. 718); Cummins, July 26, 1895; (No. 1447).

Salix macrocarpa, Nutt.
I suppose this to be very rare in the state.
Centennial Valley, June, 9, 1895 (No. 1255).

Salix monticolo, Bebb.
Infrequent; Centennial Valley, August 18, 1895 (No. 1733).

Salix rostrata, Richards, Frank. Journ. App. 753 (1823).
Our commonest *Willow*.
Laramie, June 16, 1894 (No. 244); Centennial Valley, June 9, 1895 (No. 1303); also a low mountain form from Little Sandy, August 30, 1894 (No. 1130).

Salix sp.
Only foliage but clearly enough not any of the foregoing.

Populus acuminata, Rydberg, Bull. Torr. Club, xx, 50 (1893).
Our handsomest *Cottonwood* as well as the most rapid growing of our shade trees; planted extensively in Laramie.
Fine native specimens in Whalen Conon, July 17, 1894 (No. 560). *Rydberg's Cottonwood.*

Populus angustifolia, James, Long's Exp. i, 497 (1823).
This is very frequent on the principal streams of the state, in places forming considerable bordering groves, individual trees attaining great size.
Laramie, May 16, 1894 (No. 39). *Black Cottonwood.*

Populus balsamifera, L. Sp. Pl. 1034 (1753).
This is not frequent and I have not seen it except as single trees here and there.
Dubois, August 9, 1894 (No. 749); Cummins, July 29, 1894 (No. 1547).

Populus monilifera, Ait. Hort Kew, iii, 406 (1789).

This species is used to some extent for shade purposes. If native in the state I have not yet observed it.

Populus tremuloides, Michx. Fl. Bor. Am. ii, 243 (1803).

The *Quaking Asp* of the canons and hillsides, usually only a large shrub but in some places attaining considerable size as trees.

Laramie Hills, May, 12, 1894 (No. 23); noted in scores of other places.

HYDROCHARITACEÆ.

Elodea Canadensis, Michx. Fl. Bor. Am. i, 20 (1803). *Udora Canadensis,* Nutt.

In the ponds and springs on the Fish Hatchery grounds where it has probably been introduced. October 18, 1893.

ORCHIDACEÆ.

Calypso borealis, Salisb. Parad. Lond. t. 89 (1807). *C. bulbosa,* (L.) Oakes.

This beautiful little *Orchid* is as rare here as elsewhere. The following students each found one specimen on a partly wooded hillside at the head of Pole Creek, May 25, 1894 (No. 61): Lily Boyd, Tessie Welch and Ben Bartlett

Listera convallarioides, (Sw.) Torr. Comp 320 (1826).

Very rare, only observed once; Centennial Valley, August 17, 1895 (No. 1694).

Spiranthes Romanzoffiana, Cham. Linnæa, iii, 32 (1828). *Gyrostachys Romanzoffiana,* (Cham.) MacM.

Abundant in a few localities; Centennial Valley, August 10, 1895 (No. 1663).

Habenaria gracilis, (?) Watson.

The following numbers, (420 and 1706) appear in our herbarium as *H. hyperborea,* but re-examination shows that they should probably be referred as above. Frequent in marshy places.

Chugwater Creek, July 7, 1894; Centennial Hills, August 17, 1895.

Habenaria hyperborea, R. Br. in Ait. Hort. Kew, Ed. 2, v, 203 (1813).

Specimens only from Big Wind River, August 8, 1894 (No. 725).

Habenaria obtusata, Richards. App. Frank. Journ. 750 (1823).
  Very rare; Cummins, July 30, 1895 (No. 1544).

IRIDACEÆ.

Iris Missouriensis, Nutt. Journ. Acad. Phila. vii, 58 (1834).
  Frequent on creek and river bottoms.
  Laramie, June 18, 1894 (No. 260); Centennial Valley, June 9, 1895 (No. 1268).

Sisyrinchium Bermudiana, L. Sp. Pl. 954 (1753).
  The specimens before me show the characters which were supposed to separate *S. angustifolius* and *S. mucronatum*, but the length of the spathes and the breadth of the wings are far from constant. It is well to unite them under this earlier name.
  Laramie, June 1894 (No. 239); Wind River, at Dubois, August 10, 1894 (No. 767); in various other places.

LILIACEÆ.

Streptopus amplexifolius, DC. Fl. Fran. iii, 174 (1805).
  In copses on mountain sides; Laramie Hills, June 28, 1891; Laramie Peak, August 6, 1895 (No. 1589).

Smilacina amplexicaulis, Nutt. Journ. Acad. Phil. vii, 58 (1834). *Vagnera amplexicaulis*, (Nutt.) Greene.
  Very rare; thus far only observed on Casper Mountain, July 26, 1894 (No. 640).

Smilacina stellata, Desf. Ann. Mus. Paris, ix, 52 (1807). *Vagnera stellata*, (L.) Greene.
  Frequent and abundant in the meadows bordering on all our streams.
  Horse Creek, June 9, 1894 (No. 155); Platte River, July 24, 1894 (No. 632).

Yucca angustifolia, Pursh. Fl. Am. Sept. 227 (1814). *Y. glauca*, Nutt.
  Frequent on sandy, gravelly hillsides.
  Cummins, July 27, 1895 (No. 1460). *Spanish Bayonet.*

Leucocrinum montanum, Nutt, Gray, Ann. N. Y. Lyc. iv, 110 (1837).
  Very common on the plains and foothills.
  Laramie, May 15, 1894 (No. 1226); also from Table Mountain. *White Mountain Lily.*

Allium acuminatum, Hook. Fl. Bor. Am. ii, 184, t. 146 (1840).
   On the Platte River, 1891, B. C. Buffum.
Allium brevistylum, Watson, Bot. King Exp. 350 (1871).
   Frequent in rich loam soil in copses.
   Big Horn Mountains, July 1892; Centennial Hills, August 1895 (No. 1704).
Allium cernuum, Roth. Roem. Arch. Bot. i, part 3, 40 (1798).
   Very frequent; Inyan Kara Divide, August 31, 1892; Garfield Peak, July 29, 1894 (No. 661); Laramie Peak, August 7, 1895 (No. 1633).
Allium Nuttallii, Wats. Proc. Am. Acad. xiv, 227 (1879).
   Not infrequent on gravelly river bottoms and hillsides.
   Laramie, June 19, 1894 (No. 263); Pole Creek, June 30, 1895 (No. 1382).
Allium reticulatum, Don. Mem. Wern. Soc. vi, 36 (1826-31).
   The earliest and commonest *Wild Onion* on the plains.
   Laramie, June 15, 1894 (No. 222); Pole Creek, June 29, 1895 (No. 1353).
Allium Schœnoprasum, L. Sp. Pl. 301 (1753).
   In wet places in the mountains.
   Green River, August 15, 1894 (No. 907); La Plata Mines, August 23, 1895 (No. 1788).
Fritillaria atropurpurea, Nutt. Journ. Acad. Phila. vii. 54 (1834).
   This is very rare; single specimens from the east slope of the Laramie Hills in June by B. C. Buffum, and on two occasions by Noah Wallis.
Erythronium grandiflorum, Pursh. Fl. i, 231 (1814).
   Infrequent; alpine or at least subalpine.
   Bald Mountain, August 16, 1892; La Plata Mines, August 22, 1895 (No. 1796).
Calochortus Gunnisoni, Wats. Bot. King's Exp. 348 (1871).
   Among the sage brush in the valleys.
   Sybille Creek, July 8, 1894 (No 319); Cummins, July 26, 1895 (No. 1451).
Calochortus Nuttallii, T. & G. Pac. R. R. Rep. ii, 124 (1855).
   In similar locations and nearly resembling the preceding but for the anthers and glands.
   Laramie Hills, June 18, 1891; also July 7, 1894 (No. 413).

Disporum trachycarpum, B. & H. Gen. Pl. iii, 832 (1883).
    In copses on mountain rivulets.
    Casper Mountain, July 26, 1894 (No. 609); Centennial Valley, June 9, 1895 (No. 1277).

Zygadenus elegans, Pursh. Fl. Am. Sept. 241 (1814).
    Frequent; in wet, grassy places near streams.
    Chugwater Creek, July 7, 1894 (No. 421); Cummins, July 27, 1895 (No. 1453).

Zygadenus Nuttallii, Wats. Proc. Am. Acad xiv, 279 (1879).
    Abundant on sandy plains and in the foothills.
    Laramie, June 16, 1894 (No. 254); observed in many localities.

COMMELINACEÆ.

Tradescantia Virginiana, L. Sp. Pl. 288 (1753).
    Probably confined to the eastern part of the state.
    Platte River, July 14, 1894 (No. 492).

JUNCACEÆ.*

Juncus Balticus, Willd. Berlin Mag. iii, 298 (1809).
    Very frequent in wet ground near streams.
    Laramie, June 16, 1894 (No. 243).

Juncus bufonius, L. Sp. Pl. 328 (1753).
    Frequent about spring bogs and in occasionally flooded ditches.
    Cold Springs, Fairbanks, July 13, 1894 (No. 437); Centennial Valley, August 25, 1895 (No. 1851).

Juncus longistylis, Torr. Bot. Mex. Bound. Surv. 223 (1859).
    Infrequent; Centennial Valley, August 18, 1895 (No. 1716).

Juncus Mertensianus, Bong.
    This is of frequent occurrence in the higher mountains.
    La Plata Mines, August 22, 1895 (No. 1791).

Juncus nodosus, L. Sp. Pl. Ed. 2, 466 (1762).
    Frequent; Laramie, September 1892; Little Sandy, July 1892; Wolf Creek, August 1892, by B. C. Buffum.

---

* For some further notes upon this and the order *Cyperaceæ*, see Bulletin No. 16 of this Station, by B. C. Buffum.

Juncus Parryi, Engelm. Trans. St. Louis Acad. ii, 447 (1866).
Alpine ; Teton Mountains, August 21, 1894 (No. 981); La Plata Mines, August 24, 1895 (No. 1831).

Juncus subtriflorus, (Mey.) Coville, Contrib. Natl. Herb. iv, 208 (1893).
La Plata Mines, at about 11,000 ft., August 23, 1895 (No. 1812).

Juncus tenuis, Willd. Sp. Pl. ii, 214 (1799).
Lander, August 3, 1894 (No. 699).

Juncus tenuis congesta, Engelm. Trans. St. Louis Acad. ii, 446 (1866).
Infrequent ; Laramie Peak, August 7, 1895 (No. 1631).

Juncus Torreyi, Coville.
Cold Springs, July 14, 1894 (No. 449); Teton Mountains, August 21, 1894 (No. 956).

Juncus xiphioides montanus, Engelm. Trans. St. Louis Acad. ii, 481 (1868).
Centennial Valley, August 18, 1895 (No. 1731).

Luzula spadicea parviflora, Meyer, Linnæa, xxii, 402 (1849). *Juncoides parviflorum,* (Ehrh.) Coville.
Not infrequent ; Union Pass, August 11, 1894 (No. 846).

Luzula spadicea subcongesta, Watson, Bot. Cal. ii, 202.
In boggy places and partially dried up ponds.
Centennial Valley, August 16, 1895, and June 9, 1895 (No. 1261).

Luzula spicata, DC. Fl. Fr. iii, 161 (1805). *Juncoides spicatum,* (L.) Coville.
Alpine ; Union Pass, August 11, 1894 (No. 847); noted also in the Medicine Bow Mountains

TYPHACEÆ.

Typha latifolia, L. Sp. Pl. 971 (1753).
Common in the margins of lakes and ponds.
Popo Agie River, August 1, 1894 (No. 734); abundant in the vicinity of Laramie.

ALISMACEÆ.

Alisma Plantago, L. Sp. Pl 342 (1753).
Rare ; not observed by the writer.
Dutch Creek, Sheridan County, 1892, by B. C. Buffum.

Sagittaria arifolia, (?) Nutt. in Herb. *S. variabilis minor*, Pursh.

I have some hesitancy in referring it as above, but it is the best disposition I can make of it at present.

Collected in a marshy meadow stream near Lusk, July 21, 1894 (No. 577).

Sagittaria latifolia, Willd. Sp. Pl. iv, 409 (1806).

Probably rare; Wheatland, August 11, 1894, B. C. Buffum.

NAIADACEÆ.

Triglochin maritima, L. Sp. Pl. 339 (1753).

Common in alkali marshes.

Alkali Springs, July 30, 1894 (No. 745); also from Wind River, Laramie and other localities.

Triglochin palustris, L. Sp. Pl. 388 (1753).

In similar locations; Wind River, August 8, 1894 (No. 759).

Potamogeton pectinatus, L. Sp. Pl. 127 (1753).

In the Laramie River, August 9, 1895 (No. 1668.)

CYPERACEÆ.

Eleocharis compressa, Sull. Am. Journ. Sci. lxii, 50 (1842). *E. acuminata,* (Muhl.) Nees.

As yet only from the Big Horn Mountains, July 1892, B. C. Buffum.

Eleocharis olivacea, Torr. Ann. Lyc. N. Y. iii, 300 (1836).

Not infrequent in partially submerged ground.

Laramie, June 28, 1894 (No. 289); Cold Springs, July 14, 1894 (No. 455).

Eleocharis ovata, Roem. & Schult. Syst. ii, 152 (1817).

Specimens in the World's Fair collection secured in 1892.

Scirpus atrovirens, Muhl. Gram. 43 (1817).

Very common in marshy ground about springs and ponds.

Mexican Mines, July 20, 1894 (No. 591); Muddy Creek, August 25, 1894 (No. 1109).

Scirpus lacustris, L. Sp. Pl. 48 (1753).

Rare; noted only at Cold Springs, Fairbanks, July 14, 1894 (No. 454).

Scirpus pungens, Vahl. Enum ii, 255 (1806). *S. Americanus,* Pers.
  Frequent; Cold Springs, Fairbanks, July 14, 1894 (No. 450).
Scirpus sylvaticus digynus, Boeckl. Linnæa, xxxvi, 727 (1870). *S. microcarpus,* Presl.
  Infrequent; excellent specimens from Laramie Peak, August 6, 1895 (No. 1605).
Eriophorum polystachyon, L. Sp. Pl. 52 (1753).
  Infrequent; Crazy Woman Creek, August 7, 1892, B. C. Buffum.
Carex alpina, Sw. Lilj. Sv. Fl. Ed. 2, 26, (1798).
  Probably rather rare; head of Crazy Woman Creek, August 7, 1892.
Carex athrostachya, Olney, Proc. Am. Acad. vii, 393 (1868).
  From Saratoga, August 1892.
Carex atrata, L. Sp. Pl. ii, 976 (1753).
  Infrequent; alpine; La Plata Mines, August 23, 1895 (No. 1811).
Carex aurea, Nutt. Gen. ii, 205 (1818).
  Rare; Wolf Creek, August 18, 1892, B. C. Buffum.
Carex canescens, L. Sp. Pl. 974 (1753).
  Centennial Hills, August 18, 1895 (No. 1736).
Carex deflexa media, Bailey, Mem. Torr. Club. i, 42 (1889).
  Big Sandy, July 1892, B. C. Buffum.
Carex Douglasii, Boott. Hook. Fl Bor. Am. ii, 213, t. 214 (1840).
  Secured in the Big Horn Mountains, August 8, 1892, B. C. Buffum.
Carex festiva. Dew. Am. Journ. Sci. xxix, 246 (1836).
  Very abundant and apparently throughout the state.
  Big Sandy, July, 1892; Laramie, June 28, 1894 (No. 288); also from Laramie Peak and Medicine Bow Mountains, in 1895 (Nos. 1615 and 1786).
Carex filifolia, Nutt. Gen. ii, 204 (1818).
  Infrequent; Wheatland, August, 1891.
Carex geyeri, Boott. Trans. Linn. Soc. xx, 118 (1846).
  Big Sandy, July, 1892, B. C. Buffum.
Carex Liddoni, Boott.
  Big Horn Mountains, July 1892, B. C. Buffum.
Carex marcida, Boott. Hook. Fl. Bor. Am. ii, 212, t. 213 (1840).
  In Carbon County, June 1892, B. C. Buffum.

Carex Nebraskensis, Dewey, Am. Journ. Sci. (II) xviii, 102 (1854).
  This is a very common species in wet soil everywhere, even in soils with a considerable percentage of alkali; Laramie, June 19, 1894 (No. 274); Table Mountain, June 2, 1894 (No. 130).
  Some specimens by B. C. Buffum are marked *C. Nebraskensis praevia*, Bailey, and it is not clear from the specimens nor from the literature at hand that they are different from the species.

Carex occidentalis, Bailey.
  Apparently rare; Laramie Peak, August 6, 1895 (No. 1614).

Carex Pennsylvanica, Lam. Encycl. iii, 388 (1789).
  Common on fertile hillsides among sage brush and other undershrubs.
  Crook county August, 1892; Pole Creek, May 25, 1894 (No. 69).

Carex pratensis, Drejer, Rev. Crit. Car. Bor. 24 (1841).
  Head of Crazy Woman Creek, August 8, 1892, B. C. Buffum.

Carex Raynoldsii, Dew. Am. Journ. Sci. xxxii, 39 (1837).
  Infrequent; Bald Mountain, August 15, 1892, B. C. Buffum.

Carex siccata, Dew. Am. Journ. Sci. x, 278 (1826).
  Abundant in native meadows on alkali soil.
  Laramie, June 1, 1894 (No. 65).

Carex stenophylla, Wahl. Kongl. Acad. Handl. (II.) xxiv, 142 (1803).
  Exceedingly abundant on the Laramie Plains in both dry and wet soil, the earliest *Sedge*, springing up as almost the first vegetation.
  Laramie, May 16, 1894 (No. 19); at various other times.

Carex straminea brevior, Dew.
  Rare; Lusk, July 21, 1894 (No. 586).

Carex straminea mirabilis, (Dew.) Tuck. Enum. Meth. 18 (1843).
  Specimens communicated by J. F. Lewis, of the Sheridan Experiment Farm, September 1, 1895.

Carex tenella, Schk. Riedgr. 23, Fig. 104 (1801).
  Some specimens from Wolf Creek, August 18, 1892, B. C. Buffum.

Carex Tolmiei, Boott.
  Table Mountain, June 2, 1894 (No. 100); also from the Big Horn Mountains, August 6, 1892, B. C. Buffum.

Carex utriculata, Boott. Hook. Fl. Bor. Am. ii, 221 (1840).
  Infrequent; Muddy Creek, August 25, 1894 (No. 1108).

Carex utriculata minor, Boott, l. c.
>This seems to be of more frequent occurrence than the species.
>Saratoga, 1891, by J. D. Parker; Laramie River, in Laramie county, September, 1892, B. C. Buffum.

Carex sp.
>Somewhat resembling *C. Nebraskensis*, but with culms several times as long as the leaves; spikes few, the terminal one much the longer.
>Cold Springs, at Fairbanks, July 13, 1894 (No. 447).

GRAMINEÆ.*

Andropogon provincialis, Lam. Encycl. i, 376 (1783).
>Common in the eastern part of the state.
>Wheatland, July, 1892, B. C. Buffum.

Andropogon scoparius, Michx. Fl. Bor. Am. i, 57 (1803).
>Frequent; Hartville, July 18, 1894 (No. 531); also by B. C. Buffum, 1892.

Panicum capillare, L. Sp. Pl. 58 (1753).
>Common in all parts of the state.
>Crook county, 1892; Fairbanks, July 15, 1894 (No. 553); Sheridan Experiment Farm, September 1, 1894. *Old Witch Grass.*

Panicum Crus-galli, L. Sp. Pl. 56 (1753).
>Presumably introduced but often found where such introduction would not be suspected.
>Cummins, July 29, 1895 (No. 1500); Sheridan, September 1, 1895, J. F. Lewis.

Panicum sanguinale, L. Sp. Pl. 57 (1753).
>Very generally introduced. Lander, June 1892; noted in other localities.

Panicum scoparium, Lam. Encycl. iv, 744 (1797).
>Infrequent; Whalen Canon, July 18, 1894 (No. 516).

Panicum virgatum, L. Sp Pl. 59 (1753).
>Fort Laramie, September, 1892, B. C. Buffum; Sheridan, September, 1895.

---
*The Grasses were determined in part by each of the following: Dr. W. J. Beal, Dr. F. Lamson-Scribner and Prof. B. C. Buffum. For further notes upon them see "Grasses and Forage Plants," Bulletin No. 19 of this Station, by B. C. Buffum.

Setaria viridis, Beauv. Agrost. 51 (1812). *Chamæraphis viridis,* (L.) Porter.
> A troublesome weed in cultivated ground.
> Wheatland, August 11, 1891. *Meadow Fox-tail.*

Cenchrus tribuloides, L. Sp. Pl. 1050 (1753).
> Common on the sandy plains of the Platte.
> Fort Laramie, September 5, 1892, B. C. Buffum. *Sand Bur.*

Phalaris arundinacea, L. Sp Pl. i, 55 (1753).
> Collected in 1891; no other data.

Hierochloa borealis, R. & S.
> Not frequent; Bald Mountain, August 15, 1892.

Aristida purpurea, Nutt. Trans. Am. Phil. Soc. (II.) v, 145 (1833-37). *A. fasciculata,* Torr.
> Whether specimens are typical or belong to one of the varieties of this species I am unable to say. This form is common in the state.
> Wheatland, July, 1891; Whalen Canon, July 19, 1894 (No. 539).

Stipa comata, Trin. and Rupr. Mem. Acad. St. Petersb. (VI) v, 75 (1842).
> Frequent on the plains and in dry valleys; when mature an annoying and worthless grass.
> Laramie, June 17, 1891; June 28, 1894 (No. 284).

Stipa spartea, Trin. Mem. Acad. St. Petersb. ser. 6, i, 82 (1829).
> On the plains of the eastern part of the state, 1891.

Stipa viridula, Trin. Mem. Acad. St. Petersb. ser. 6, v, 39 (1836).
> Big Sandy, Fremont county, July, 1892.

Oryzopsis exigua, Thurb. Bot. Wilkes Exp. 481.
> Hitherto, it seems, reported only from Oregon and Washington.
> Big Sandy, in the Wind River Mountains, July 22, 1892, B. C. Buffum.

Oryzopsis menbranacea, Vasey, Grasses S. W. part 2, t. 10 (1891).
> Very abundant on the plains and in the foothills.
> University Campus, September 16, 1894 (No. 1137); noted in many other places. *Mountain Rice.*

Muhlenbergia comata, Benth. Vasey's Cat. Grasses U. S. 39 (1885).
> Spring Creek, Big Horn Mountains, August 5, 1892.

Muhlenbergia dumosa, Scrib. in Vasey, Contr. Nat. Herb. iii, 71.
   Frequent in the eastern part of the state.
   Wheatland, August 16, 1891.

Muhlenbergia glomerata, Trin. Unifl. 191 (1824). *M. racemosa,*
(Michx.) B. S. P.
   From Wolf Creek, Rawhide Creek and Fort Laramie, 1892, B.
C. Buffum.

Muhlenbergia gracilis breviaristata, Vasey, Rothr. in Wheeler's
Rep. vi, 284.
   Plumbago Canon, July 1891.

Phleum alpinum, L. Sp. Pl. 59 (1753).
   Common in the alpine region of all our mountain ranges.
   Union Pass, August 10, 1894 (No. 863); La Plata, August 22,
1895 (No. 1781). *Wild Timothy.*

Alopecurus geniculatus fulvus, (J. E. Smith) Scrib. List. Pterid &
Sperm. 38 (1894).
   Frequent; Whalen Canon, July 19, 1894 (No. 544); Centennial
Valley, August 18, 1895 (No. 1721).

Sporobolus airoides, Torr. Pac. R. R. Rep. vii, part 3, 21 (1856).
   Fort Washakie Hot Springs, July, 1892, B. C. Buffum.

Sporobolus asperifolius, Thurb. in Wats. Bot. Cal. ii, 269 (1880).
   Probably infrequent; Cheyenne, August, 1891.

Sporobolus cryptandrus, Gray, Man. 576 (1848).
   Rather frequent, on dry ridges.
   Wheatland, September 8, 1892; Fairbanks, July 13, 1894 (No.
435).

Sporobolus cuspidatus, Scribn. Bull. Torr. Club, x, 63 (1882). *S. brevifolius,* (Nutt.) Scribn.
   Fremont county, July, 1892.

Sporobolus heterolepis, Gray, Man. 576 (1848).
   Crook county, August, 1892.

Agrostis alba, L. Sp. Pl. 63 (1753).
   Common; Crook county, 1892; Laramie, October 6, 1894 (No.
1172).

Agrostis asperifolia, Trin. Mem. Acad. St. Petersb. Sci. Nat. ser. 6,
318 (1845).
   Infrequent; Centennial Valley, August 18, 1895 (No. 1719).

—14

Agrostis exarata, Trin. Unifl. 207 (1824).
  Head of Rawhide Creek, August 1892.
Agrostis humilis, Vasey, Bull. Torr. Club, x, 21.
  Alpine, in wet soil; La Platta Mines, August 23, 1895 (No. 1814).
Agrostis scabra, Willd. Sp. Pl. i, 370 (1798).
  Frequent; Big Sandy, July, 1892; Albany county, September, 1891.
Calamagrostis Canadensis, Beauv. Agrost. 15 (1812).
  Wind River Mountains, July, 1892.
Calamagrostis confinis, Nutt. Gen. i, 47 (1818).
  Common; Centennial Valley, September 7, 1891; Laramie, September 30, 1894 (No. 1179).
Calamagrostis longifolia, Hook. Fl. Bor. Am. 241 (1840). *Calamovilfa longifolia*, (Hook.) Hack.
  Frequent; Lander, July, 1892; Whalen Canon, July 9, 1894 (No. 536).
Calamagrostis Montanensis, Scribn.
  Sheridan county, August, 1892.
Calamagrotis neglecta, (Ehrh.) Gaertn. Fl. Wett. i. 94 (1799).
  Frequent; Big Horn Mountains, August, 1892; Orin Junction, August 14, 1891.
Calamagrostis pallida, Vasey & Scribn.
  Eagle Rock Canon, September 22, 1891; also from Carbon county.
Calamagrostis purpurascens, R. Br., in Rich. Bot. App. Frank. Voy. 3.
  Infrequent; Laramie Peak, August 7, 1895 (No. 1627).
Deschampsia cæspitosa, Beauv. Agrost. 160, t. 18, f. 3 (1812).
  Common in the mountains; Big Sandy, July, 1892; La Plata Mines, August 23, 1895 (No. 1815).
Deschampsia cæspitosa arctica, Vasey.
  Rare; Laramie Hills, June 12, 1894 (No. 238).
Trisetum subspicatum, Beauv. Agrost. 180 (1812).
  Very abundant in the mountains; Big Sandy, July, 1892; La Plata Mines, August 22, 1895 (No. 1783).
Trisetum subspicatum molle, Gray, Man. Ed. 2, 572 (1856).
  La Plata, Medicine Bow Mountains, August 23, 1895 (No. 1807).
Trisetum Wolfii, Vasey.
  Fort Laramie, September 6, 1892, B. C. Buffum.

Danthonia Californica, Boland. Proc. Cal. Acad. ii, 182 (1862-63).
   Sheridan county, August, 1892.
Danthonia intermedia, Vasey.
   Big Sandy, Fremont county, July, 1892.
Danthonia unispicata, Munro.   *D. Californica unispicata,* Thurb. Bot. Cal. ii, 294 (1880).
   Apparently rare; Laramie Peak, August 7, 1895 (No. 1630).
Spartina cynosuroides, Willd. Enum. Pl. 80 (1809).
   Common and widely distributed; Inyan Kara Divide, August 30, 1892; C. Y. Ranch, on Big Muddy, July 23, 1894 (No. 599).
Schedonnardus Texanus, Steud. Syn. Pl. Gram. 146 (1855).   *S. paniculatus,* (Nutt.) Scribn.
   Abundant on the plains of the eastern part of the state. Orin Junction, August, 1891.
Bouteloua oligostachya, Torr. Gray, Man. Ed. 2, 553 (1856).
   A very common grass on the plains.
   Wheatland, June 11, 1891 ; noted in many localities.
Bouteloua racemosa, Lag. Var. Clenc. y Litter. 2 : Part 4, 141 (1805).   *B. curtipendula,* (Michx.) Torr.
   This, like the preceding, is very abundant in many localities.
   Inyan Kara Divide, August 30, 1892; Hartville, July 18, 1894 (No. 530).
Beckmannia erucæformis, Host. Gram. Austr. iii, 5 (1805).
   In swampy ground along our streams.
   Big Sandy, July, 1892; Union Pass, August 11, 1894 (No. 828); Cummins, July 30, 1895 (No. 1533).
Buchloe dactyloides, Engelm. Trans. St. Louis Acad. i, 432 (1859).   *Bulbilis dactyloides,* (Nutt.) Raf.
   This is still found in considerable abundance on some of our plains, but it seems that it is gradually being killed out and replaced by other grasses.
   Wheatland, September, 1892, and noted in many parts of the state.   *Buffalo Grass.*
Munroa squarrosa, Torr. Pac. R. R. Rep. iv, 158 (1856).
   Noted only in the eastern part of the state.
   Wheatland, August, 1891 ; Blue Grass Creek, July 8, 1894 (No. 372).

Phragmites communis, Trin. Fund. Agrost. 134 (1820). *P. Phragmites,* (L.) Karst.

Not noted by the writer, but fine specimens by B. C. Buffum, from Little Muddy Creek, Casper, August 12, 1891. *Reed Grass.*

Eragrostis major, Host. Gram. Austr. 4: 14, t. 21 (1809).

Sparingly introduced about Wheatland and probably in other localities: September, 1892, B. C. Buffum.

Eatonia obtusata, Gray, Man. Ed. 2, 558 (1856).

Frequent; Crook county, August, 1892; Fairbanks, July 14, 1894 (No. 459).

Koeleria cristata, Pers. Syn. i. 97 (1812).

Common everywhere; Johnson county, August, 1892; Laramie, June 28, 1894 (No. 283), and noted in numerous localities.

Catabrosa aquatica, Beauv. Agrost. 157, t. 19, f. 8 (1812).

Throughout the state; Crook Creek, June 28, 1894; Cottonwood Canon, August 4, 1895 (No. 1567).

Melica bulbosa, Geyer.

Infrequent; Bald Mountain, August 16, 1892.

Distichlis maritima, Raf. Journ. Phys. lxxxix. 104 (1818). *D. spicata,* (L.) Greene.

Frequent on wet alkali flats.

Big Wind River, August 1, 1892; Alkali Springs, July 31, 1894 (No. 659).

Poa alpina, L. Sp. Pl. 67 (1753).

Rare, as well as rarely beautiful; Bald Mountain, August 15, 1892.

Poa andina, Nutt. Wats. Bot. King Exp. 388 (1871). *P. arida,* Vasey.

Common on dry slopes; Middle Pass, June 21, 1892; Laramie, June 5, 1895 (No. 1245).

Poa Californica, Vasey, Cat. Grasses, U. S. 81 (1885). *P. Fendleriana.* (Steud.) Vasey.

Infrequent; Big Sandy, July, 1892.

Poa cæsia strictior, Gray.

Laramie Hills, June, 1892, B. C. Buffum.

Poa cuspidata, Vasey.

Carbon county, June, 1892; La Plata Mines, August 22, 1895 (No. 1782).

Poa lævis, Vasey.

Carbon county, June, 1892.

Poa nemoralis, L. Sp. Pl. 69 (1753).
    This is of frequent occurrence; Big Sandy, July, 1892; Laramie, June 19, 1894 (No. 272).
Poa Nevadensis, Vasey, Bull. Torr. Club, x, 66 (1883).
    I am unable to say whether this is frequent or not. I notice it mentioned by others as a valuable forage grass. Specimens at hand only from Big Sandy, July, 1892.
Poa occidentalis, Vasey.
    This fine species is common on the streams near Laramie; Fish Hatchery, July, 1891; Laramie River, June 22, 1895 (No. 1317).
Poa pratensis, L. Sp. Pl. 67 (1753).
    Collected in 1892 for the World's Fair exhibit; common. *Blue Grass.*
Poa reflexa, Vasey & Scribn.
    An alpine or subalpine species.
    Bald Mountain, August 16, 1892; La Plata Mines, August 24, 1895 (No. 1837).
Poa rupestris, Vasey.
    Rare; noted but once at 11,000 ft.; La Plata Mines, August 23, 1895 (No. 1813).
Poa serotina, Ehrh. Beitr. vi, 83 (1793). *Poa flava,* L.
    Cheyenne, August 11, 1891. *False Red Top.*
Poa tenuifolia, Nutt. Buckley, Proc. Acad. Phila. 1862, 96 (1862).
    Very common throughout the state.
    Big Sandy, July, 1892; Laramie Hills, May 23, 1894 (No. 49). *Bunch Grass.*
Poa Vaseyana, Scribn.
    This fine species was found in abundance by Prof. Buffum, on the Big Sandy, July 1892. He sent it for determination to Dr. Beal, who pronounced it probably new. Whether specimens were later sent to the Department of Agriculture or the same had been collected by Dr. Vasey I am unable to say, but subsequently it received the above name.
Poa sp.
    Of this, Dr. F. Lamson Scribner writes: "I am not prepared to name it. It has some characters in common with *Poa occidentalis,* Vasey, but it is apparently distinct from that."
Glyceria arundinacea, Kunth.
    On Spring Creek, Big Horn Mountains, August 5, 1892.

Glyceria grandis, Wats. Gray. Man. Ed. 6, 667 (1890). *Panicularia aquatica,* (L.) Kuntze.
    Common in partially submerged meadow lands.
    Big Horn Valley, July, 1892; Cummins, July 30, 1895 (No. 1534).

Glyceria pauciflora, Presl. Rel. Haenk. i, 257 (1830). *Panicularia pauciflora,* (Presl.) Kuntze.
    Infrequent; Centennial Valley, August 18, 1895 (No. 1720).

Festuca confinis, Vasey.
    Big Horn Mountains, August, 1892; Carbon county, June 11, 1892; Pole Creek June 2, 1894 (No. 161); Wallace Creek, July 29, 1894 (No. 667).

Festuca gracillima, Hook.
    Big Sandy, Wind River Mountains, July, 1892.

Festuca ovina, L. Sp. Pl. 73 (1753).
    From Laramie county, August, 1891.

Festuca ovina brevifolia, Chlor. Melv. 289 (1823).
    Big Sandy, July, 1892.

Festuca rubra, L. Sp. Pl. 74 (1753).
    Wind River Mountains, July, 1892.

Festuca tenella, Willd. Enum i, 113 (1809). *F. octoflora,* Walt. Fl. Car. 81.
    Wheatland, 1892, by M. R. Johnson.

Bromus breviaristatus, (Hook.) Buckl. Proc. Acad. Phila. 1862, 98 (1862).
    Wolf Creek, July, 1892.

Bromus ciliatus, L. Sp. Pl. 76 (1753).
    Inyan Kara Divide, August 30, 1892.

Bromus Hookerianus, Thurb.
    Centennial Valley, September 7, 1894.

Bromus Kalmii occidentalis, Vasey.
    Union Pass, August 11, 1894 (No. 821); Sand Creek, Fremont County, August 27, 1894 (No. 1105).

Bromus Pumpellianus, Scribn.
    Big Wind River, July 1892.

Agropyrum caninum, R. & S. Syst. ii, 756 (1817).
    Specimens in the World's Fair collection, 1892.

Agropyrum divergens, Nees. Steud. Syn. Gram. 347 (1855).
    Frequent in the northern part of the state.
    Lander, July 1892; Union Pass, August 11, 1894 (No. 820).

Agropyrum glaucum, R. & S. Syst. Veg. ii, 752 (1817). *A. repens glaucum,* (Desf.) Scribn.

One of the most valuable of the native hay grasses, producing heavy crops under judicious irrigation. It is essentially an upland grass and over-irrigation will soon destroy it.

Lander, July 1892 ; Laramie and other places at various times.

Agropyrum unilaterale, Cassidy, Bull. Col. Exp. Station xii, 63 (1890) *A. caninum unilaterale* (Cassidy) Vasey.

Sheridan, August 18, 1892.

Agropyrum violaceum, Vasey, Grasses U. S. Spec. Rep. No. 63, 45. A valuable forage plant; Big Sandy, July 1892.

Hordeum jubatum, L. Sp. Pl. 85 (1753).

The worst weed in the state; a positive pest in the hay fields. Its extermination should receive every encouragement.

Hartville, July 16, 1894 (No. 558), and noted in numerous places.

Hordeum nodosum, L. Sp. Pl. Ed. 2, 126 (1762).

Infrequent; Big Horn Mountains, August 4, 1892.

Hordeum pusillum, Nutt. Gen. i, 87 (1818).

Infrequent; Platte River, Fairbanks, July 11, 1894 (No. 426).

Elymus Canadensis, L. Sp. Pl. 83 (1753).

Frequent but hardly common; Whalen Canon, July 18, 1894 (No. 552), and by B. C. Buffum in 1892.

Elymus condensatus, Presl. Rel. Haenk. i, 265 (1830).

Common along streams in the eastern part of the state Wheatland, 1892.

Elymus sitanion, Schult. Mant. ii, 426 (1824). *E. elymoides,* (Raf.) Swezey.

On mountain slopes at high altitudes.

Union Peak, August 13, 1894 (No. 1021); Laramie Peak, August 6, 1895 (No. 1602).

Elymus Virginicus, L. Sp. Pl. 84 (1753).

Not common ; Prairie Dog, August 8, 1892.

CONIFERÆ.

Pinus flexilis, James, Long Exp. ii, 35 (1823).

This is common in our mountain ranges.

Laramie Hills, May 12, 1894 (No. 18); Cummins, July 29, 1895 (No. 1501). *Rocky Mountain White Pine.*

Pinus Murrayana, Balfour, Jeffr. Rep. Oreg. Exp. (1853).

Noted in the foothills in the Laramie and Medicine Bow ranges, where it is of frequent occurrence along streams.

Our specimens from Pole Creek, May 12, 1894 (No. 12); May 18, 1895 (No. 1213). *Lodge Pole Pine.*

Pinus ponderosa scopulorum, Engelm. Wats. Bot. Cal. ii, 126 (1880).

This forms a somewhat scattering growth on the higher more exposed ridges in the Laramie range, and less conspicuously so in the other ranges visited.

Laramie Hills, May 12, 1894 (No. 17). *Rocky Mountain Yellow Pine.*

Picea Engelmanni, Engelm.

This forms a considerable proportion of the forest growth in the Medicine Bow and probably in the other ranges of our state. Attaining its most luxuriant growth at about 9,000 ft., it is the sole survivor of the trees at timber line and there becomes reduced and spreading-prostrate. Said to be the most valuable of our trees for lumber.

Union Pass, August 13, 1894 (No. 1014); La Plata Mines, August 24, 1895 (No. 1841). *Engelmann's Spruce.*

Picea pungens, Engelm.

This is much less common and usually occurs along streams in the wooded foothills. It is considered the most beautiful of our Spruces and is well worthy of the high esteem in which it is held as an ornamental tree.

Laramie Hills, May 12, 1894 (No. 16); Cummins, July 30, 1895 (No. 1549). *Blue Spruce. Balsam.*

Pseudotsuga Douglasii, Carr.

The largest of our forest trees, attaining a remarkable size in the lower altitudes of our mountain ranges.

Laramie Hills, May 12, 1894 (No. 14); April 1895 (No. 1208); *Douglas Spruce.*

Juniperus communis, L. Sp. Pl. 1040 (1753).

Very rare; Cummins, July 28, 1895 (No. 1481).

Juniperus communis alpina, Gaud. Fl. Helv. vi, 301 (1830). *J. nana,* Willd.

This is abundant on hillsides at all altitudes.

Laramie Hills, May 4, 1894 (No. 11); Little Sandy, August 30,

1894 (No. 1128). *Prickly Juniper.*

Juniperus Sabina procumbens, Pursh. Fl. Am. Sept. 647 (1814). *J. Sabina,* L.

Rare; observed only on the alpine summits of the Medicine Bow Mountains. La Plata, August 24, 1895 (No. 1834).

Juniperus Virginiana, L. Sp. Pl. 1039 (1753).

This is frequent, varying within our range from a prostrate, scraggly shrub to a large tree. Incorrectly called *Red Cedar.*

Laramie Hills, April 4, 1894 (No. 1). *Virginia Juniper.*

EQUISETACEÆ.

Equisetum arvense, L. Sp. Pl. 1061 (1753).

Frequent on water courses.

Pole Creek, May 25, 1894 (No. 77); July 2, 1895 (No 1411).

Equisetum arvense alpestre, Wahl.

Not common; occurring usually on abrupt wet creek banks.

Big Wind River, August 5, 1894 (No. 706).

Equisetum hiemale, L. Sp. Pl. 1062 (1753).

Common on sandy river bottoms.

Laramie River, June 19, 1894 (No. 275); Laramie Peak, August 8, 1895 (No. 1640). *Scouring Rush.*

Equisetum lævigatum, A. Br. Engelm. Am. Journ Sci. xlvi, 87 (1844).

Infrequent; C. Y. Ranch on the Big Muddy, July 23, 1894 (No. 604).

Equisetum variegatum, Schleich. Cat. Pl. Helv. 27 (1807).

Fish Hatchery at Laramie, July 1891; Pole Creek, June 28, 1895 (No. 1355).

FILICES.

Asplenium Filix-fœmina, (L.) Bernh. Schrad. Neues. Journ. Bot. i, part 2, 26 (1806).

Infrequent; Jackson's Hole, August 21, 1894 (No. 940).

Cheilanthes lanuginosa, Nutt. Hook. Sp. Fil. ii, 99 (1858). *C. gracilis,* Mett.

In dry, rocky, cliffs; Laramie, 1891; Platte Canon, July 13, 1894 (No. 442).

Chryptogramma acrostichoides, R. Br App. Frank. Journ. 767 (1823).

Teton Mountains, August 21, 1894 (No. 956).

Cystopteris fragilis, (L.) Bernh. Schrad. Neues. Journ. Bot. i, part 2, 27 (1806).
    Common; Pole Creek, June 2, 1894 (No. 97); Mexican Mines, July 20, 1894 (No. 587).
Notholæna sinuata, Kaulf.
    A specimen by B. C. Buffum, July 8, 1892; no other data.
Pteris aquilina, L. Sp. Pl. 1075 (1753).
    Teton Mountains, August 21, 1894; Laramie Peak, August 6, 1895 (No. 1601).
Woodsia Oregana, Eaton, Can. Nat. ii, 90 (1865).
    Centennial Hills, August 16, 1895 (No. 1682).
Woodsia scopulina, Eaton, l. c.
    Much more frequent than the preceding; in crevices and on rocky ledges throughout our range.
    Teton Mountains, August 21, 1894 (No. 951); Laramie Peak, August 6, 1895 (No. 1594).

SELAGINELLACEÆ.

Selaginella rupestris, Spring, in Mart. Fl. Bras. i, part 2, 118 (1840).
    On a dry, naked ridge near table Mountain, June 28, 1895 (No. 1345).

MUSCI.[*]

Ceratodon purpureus, Brid. Bryol. Univ. i, 480 (1826).
    Head of Pole Creek, May 18, 1895 (No. 1211 in part).
Desmatodon latifolius glacialis, Schimp. Syn. 157.
    Nearly alpine; La Plata Mines, August 24, 1895 (No. 1835).
Desmatodon Porteri, James, Aust. Musc. Appal. n. 123.
    Cummins, July 30, 1895 (No. 1538 in part).
Barbula mucronifolia, Bruch. & Schimp. Mon. xxxviii, t. 23 (1842).
    Cummins July 30, 1895 (No. 1538 in part).
Philonotis fontana, Brid. Bryol. Univ. ii, 18 (1827).
    Centennial Valley, August 18, 1895 (No. 1756); LaPlata Mines, August 22, 1895 (No. 1800).
Aulacomnium papillosum, Lesq. & James, Man. Moss. N. A. 253.
    Centennial Valley, August 19, 1895 (No. 1746).

[*] The Musci were determined and arranged by Prof. J. M. Holzinger.

Leptobryum pyriforme, Schimp. Syn. Musc. Eur. 390 (1876).
  Centennial Valley, August 18, 1895 (No. 1717).
Webera albicans, Schimp. Coroll. 67.
  Pole Creek, near Table Mountain, July 1, 1895 (No. 1285).
Webera elongata, Schwægr. Spec. Musc. 48.
  Centennial Valley, August 18, 1895 (No. 1723 in part).
Webera sp.
  Specimens sterile, but different from the preceding.
  Centennial Valley, August 19, 1895 (No. 1745)
Bryum cæspiticium, L. Sp. Pl. ii, 1121 (1753).
  Head of Pole Creek, May 18, 1895 (No. 1211 in part).
Bryum cirrhatum, Hoppe & Hornsch. Fl. 90 (1819). var. ?
  Cummins, July 30, 1895 (No. 1538 in part).
Bryum intermedium, Brid. Musc. Recent Suppl. iv, 120.
  Centennial Valley, June 8, 1895 (No. 1263).
Bryum pseudotriquetrum, Schwægr. Suppl. i, 2, 110.
  Centennial Valley, June 8, 1895 (No. 1259).
Mnium subglobosum, Bruch. & Schimp. Bryol. Eur. t. 388.
  Centennial Valley, August 19, 1895 (No. 1744).
Timmia Austriaca, Hedw. Sp. Musc. 176, t. 42 (1801).
  Laramie Peak, August 8, 1895 (No. 1645).
Polytrichum juniperinum alpinum, Schimp. Syn. 447.
  La Plata Mines, August 24, 1895 (No. 1830).
Polytrichum piliferum, Schreb. Spicil. Fl. Lips. 74.
  La Plata mines, August 24, 1895 (No. 1769).
Climacium Americanum, Brid. Musc. Recent Suppl. ii, 45.
  Centennial Valley, August 19, 1895 (No. 1724).
Pseudoleskea oliogoclada, Vindb.
  Centennial Valley, August 18, 1895 (No. 1734).
Thuidium Blandovii, Bruch. & Schimp. Bryol. Eur. t. 486.
  Centennial Valley, August 19, 1895 (No. 1746 in part).
Brachythecium acutum, Sulliv. Icon Musc. Suppl. 99, t. 75.
  Centennial Valley, August 17, 1895 (No. 1698).
Brachythecium rivulare, Bruch & Schimp. Bryol. Eur. t. 543.
  Laramie Peak, August 7, 1895 (No. 1622).
Hypnum commutatum, Hedw. Musc. Frond. iv, 68, t. 24.
  Gros Ventre River, August 16, 1894 (No. 1088).

Hypnum plicatile, Lesq. & James. Man. Moss. N. A. 394.
Cummins, July 29, 1895 (No. 1507).

MARCHANTIACEÆ.

Marchantia polymorpha. L.
Frequent on wet banks ; Green River, August 25, 1894 (No. 1005).
Centennial Valley, August 18, 1895 (No. 1748).

ALGÆ.

The Algæ have received no attention so far as specific determination is concerned. In the collection of material for class use in the laboratory it has been found that quite a large number of genera are represented, some of them by a number of species. Among these *Spirogyra*, *Zygnema* and *Vaucheria* may be named. *Diatomaceæ* are everywhere but the *Desmidiaceæ* are not so well represented.

Among the larger forms the two following are conspicuous in the ponds at the city springs :

Chara fœtida, A. Br.
It forms a dense growth over the bottom of the ponds, in places reaching a foot or more in height.

Batrachospermum gelatinosum, (L.) A. F. Woods, Rep. Bot. Surv. Neb. iii, 6 (1894).
Adherent to stones in running water.

Chara sp.
A rather unusual *Chara* was collected in a pool in the mouth of an extinct geyser pan on Warm Spring Creek. It is in the hands of Mr. J. W. Blankinship for determination ; August 9, 1894 (No. 796).

FUNGI.

The following Fungi have been determined by Mr. J. B. Ellis. They include only incidental "pickups" in the field. Those of economic importance that we have had to deal with on the Experiment Farm are not included.

Æcidium abundans, Pk.
Cummins, July 29, 1895 (No. 1498). On *Symphoricarpos eriophilus*.

Æcidium monoicum, Pk.
 Laramie Hills, May 23, 1894 (No. 50). On *Sisymbrium linifolium*.
Æcidium Œnothera, Mont.
 Pole Creek, June 2, 1894 (No. 133). On *Œnothera brachycarpa*.
Uromyces Junci, (Schw.) Tul.
 Laramie, December, 1895 (No. 1207). On *Juncus sp.*
Erysiphe cichoracearum, DC.
 Cummins, July 29, 1895 (No. 1516). On *Hydrophyllum occidentale*.
Melampsora farinosa, (Pers.) Schrœt.
 Cummins, July 29, 1895 (No. 1520); Centennial Valley, August 25, 1895 (No. 1864). Frequent on various *Willows*.
Phragmidium subcorticum, Schrank.
 Cummins July 29, 1895 (No. 1499). On *Rosa blanda* (?).
Ramularia sidalcea, E. & E.
 Cummins, July 29, 1895 (No. 1468). On *Sidalcea candida*. Mr. Ellis writes of this as follows: "I have seen it but once before. It was from British Columbia, sent by Dr. Macoun."

LICHENES.

 The following are a few of our commoner *Lichens*:
Claydonia pyxidata, Fr. Centennial Valley, August 18, 1895 (No. 1749).
Evernia vulpina, Ack. Centennial Hills, August 17, 1895 (No. 1699).
Parmelia conspersa, Ehrh. Pole Creek, July 1, 1895 (No. 1387).
Parmelia molliuscula, Ack. Laramie, July 23, 1895 (No. 1429).
Peltigera aphthosa, Hoffm. Laramie Peak, August 8, 1895 (No. 1644).
Peltigera canina, Hoffm. Cummins, July 29, 1895 (No. 1508).
Rinodina turfacea, Koerb. Centennial Hills, August 17, 1895 (No. 1713).

Appendix to List of Species.

*Arabis Drummondii, Gray, Proc. Am. Acad. vi, 187.
    On fertile hillsides in the mountains; not frequent.
    Union Pass, August 10, 1894 (No. 875); Centennial Hills, June 7, 1895 (No. 1248).

Arabis hirsuta, (L.) Scop. Fl. Carn. Ed. 2, ii, 30 (1772).
    In sandy valleys, sometimes among the sage brush; frequent.
    Laramie Hills, June 7, 1894 (No. 181); Union Pass, August 13, 1894 (No. 1074); Pole Creek, July 1, 1895 (No. 1394).

Arabis Holbœllii, Hornem. Fl. Dan. t. 1879 (1827).
    On a dry, stony sandbar of the Laramie River, Cummins, July 30, 1895 (No. 1551).

Arabis Holbœllii Fendleri, Wats. Syn. Fl. i, 164 (1895).
    Noted only in the sand beds of the stony foothills of the Laramie range; May 16, 1894 (No. 32).

Arabis Lemmoni, Wats. Proc. Am. Acad. xxii, 467 (1887).
    Among the sage brush on the plains; infrequent.
    Laramie, May, 23, 1894 (No. 56).

Arabis Lyalli, Watson, Proc. Am. Acad. xi, 122.
    Alpine; Teton Mountains, August 21, 1894 (No. 1007).

Arabis Nuttallii, Robinson, Syn. Fl. i, 160 (1895).
    In valleys but on dry ground.
    Horse Creek, June 9, 1894 (No. 218); Pole Creek, June 28, 1895 (No. 1359).

Arabis perfoliata, Lam. Encycl. i, 219 (1788).
    Rare; noted but once; Laramie Peak, August 7, 1895 (No. 1628).

Thelypodium integrifolium, Endl. in Walp. Rep. i, 172 (1842).
    Very frequent, especially on saline plains.
    Lusk, July 21, 1894 (No. 574); Dubois, August 9, 1894 (No. 747); Laramie, August 10, 1895 (No. 1663).

* Determinations in this genus by Dr. B. L. Robinson.

Thelypodium sagittatum, Endl. l. c.
>Widely distributed, but only scattering plants.
>Wheatland, June 18, 1891; Pole Creek, June 2, 1894 (No. 112); Bacon Creek, August 15, 1894 (No. 922).

Sisymbrium virgatum, Nutt. in T. & G. Fl. i, 93.
>On sandy ground among the sage brush.
>Laramie Plains, June 9, 1895 (No. 1299).

Draba nemorosa, L. Sp. Pl. 643 (1753).
>Frequent in wet loam soil in valleys; variable, some of our specimens are the variety *hebecarpa*, Lindb.
>Pole Creek, June 2, 1894 (No. 153); Union Pass, August 11, 1894 (No. 854); Centennial Valley, June 9, 1895 (No. 1254); at other times and places.

Arenaria Nuttallii, Pax in Engler, Jahresb. xviii, 30.
>Infrequent; noted only at Garfield Peak, July 29, 1894 (No. 675).

Astragalus bodini, Sheld.
>Very abundant in meadow lands in the Centennial Valley; August 25, 1895 (No. 1855).

Astragalus leucopis, Torr. Mex. Bound. Surv. 56, t. 16.
>Rare; specimens from the eastern part of the state by B. C. Buffum in 1892.

Astragalus Parryi, Gray, Am. Journ. Sci. ii, 33.
>Frequent on gravelly hillsides; Pole Creek, June 2, 1894 (No. 101); Centennial Valley, June 9, 1895 (No. 1298).

CORRECTIONS:—

On page 63, line 3 from bottom, read Engelmann's Spruce for Douglas Spruce.

In the *Astragali* some of the names as given in the list of *Pteridophyta* and *Spermaphyta*, which it was intended to give as synonyms, were inadvertently omitted.

New Species and Varieties.

The succeeding list includes the new species, varieties and names as published in this report.

	PAGE
Aquilegia cærulea alpina, n. var.	78
Aquilegia Laramiensis, n. sp.	78
Aconitum Columbianum ochroleucum, n. var.	79
Thlaspi alpestre glaucum, n. var.	84
Trifolium longipes reflexum, n. var.	94
Oxytropis Lamberti ochroleuca, n. var.	98
Potentilla pinnatisecta, n. sp.	104
Erigeron uniflorus melanocephalus, n. var.	131
Hymenopappus liguletlorus, n. sp.	135
Actinella glabra, (Nutt.)	136
Artemisia Ludoviciana integrifolia, n. var.	138
Senecio Douglasii, (Some forms of)	141
Senecio lugens megalocephalus, n. var.	142
Hieracium gracile minimum, n. var.	144
Androsace septentrionalis subumbellata, n. var.	149
Mertensia lanceolata viridis, n. var.	158

The following have recently been published as new from this state:

Ranunculus eximus, Greene.	77
Tissa sparsiflora, Greene	88
Amelanchier pumila, (Nutt.) Greene	106
Mitella trifida integripetala, Rose	107
Mentzelia Nelsonii, Greene	113
Chrysothamnus collinus, Greene, Pitt. iii, part 13, 24 (1896)	122
Chrysothamnus linifolius, Greene, l. c.	123
Allocarya Nelsonii, Greene	156
Eriogonum subalpinum, Greene, Pitt. iii, part 13, 18 (1896)	174

Lists of Plants Reported by Other Collectors.

In the succeeding lists such plants are given as have been reported by others from this state but are as yet unrepresented in our herbarium. It is intended to exclude all names that are merely synonyms of those of our own list, but even with greatest care I fear some will get in. Some, while not synonyms, are names of plants that very doubtfully belong to this region at all. On the other hand these lists, undoubtedly, do not represent all that have been reported from this state, but are intended to be complete with respect to the literature at hand. Such references as could not somewhat satisfactorily be verified are excluded, probably too many.

FROM TORREY'S REPORT ON THE PLANTS OF FREMONT'S EXPEDITION, 1842.

Astragalus tridactylicus, Gray, as *Phaca digitata*, Torr. Little Sandy, August 8.
Potentilla Pennsylvanica glabrata, Watson, as *P. sericea glabrata*, Lehm. Sweetwater River, August 4-15.
Sedum rhodiola, DC. Wind River Mountains, August 12-17.
Symphoricarpos vulgaris, Michx. Wind River Mountains, August 13-14.
Aster Novae-Angliae, L. Wind River Mountains, August 18.
Aster andinus, Nutt. Wind River Mountains, August 16.
Erigeron salsuginosus glacialis, Gray, as *Aster glacialis*, Nutt. Wind River Mountains, August 16.
Solidago nemoralis incana, Gray, as *S. incana*, T. & G. Sweetwater River.
Helianthus petiolaris, Nutt. Laramie Hills, July 26.
Helianthus Maximiliana, Schrader. Laramie Hills, July 26.
Hymenopappus corymbosus, T. & G. Upper Platte, August 26.
Vaccinum myrtilloides, Hook. Wind River Mountains, August 15.
Dodecatheon Meadia latilobum, Gray, as *D. dentatum*, Hook., Wind River Mountains, August 13-16.
Phlox muscoides, Nutt. Wind River Mountains, August 15.
Gentiana arctophila, Griseb. Sweetwater River, August 4.
Gentiana linearis, Froel. as *G. pneumonanthe*, L. Laramie River, July 12.

Habenaria leucophaea, Gray, as *Plantanthera leucophaea*, Lindl. Laramie (Black) Hills, July 27.
Spiranthes cernua, Rich. Sweetwater River, August 6.

FROM COULTER'S REPORT ON THE BOTANY OF THE HAYDEN SURVEY, 1872.

Clematis alpina occidentalis, Gray, Teton Mountains, July.
Ranunculus Nelsonii, Gray, Yellowstone Lake.
Aquilegia flavescens, Wats. Yellowstone Lake, August.
Delphinium Menziesii, DC. Teton Mountains, July.
Aconitum Fischeri, Wats. as *A. nasutum*, Hook. Yellowstone Lake, August.
Nuphar advena, Ait. Yellowstone Park, August.
Dicentra uniflora, Kellogg. Teton Mountains, August.
Cardamine oligosperma, Nutt. Teton Mountains, July.
Draba aurea, Vahl. Teton Mountains, July.
Draba nivalis, Lilj. as *D. Stellata*, Jacq. Teton Mountains, July.
Lychnis Drummondii, Wats. Yellowstone Park, July.
Arenaria pungens, Nutt. Teton Mountains, July.
Arenaria verna, L. Teton Mountains, July.
Arenaria laterifolia, L. Teton Mountains, July.
Sagina Linnaei, Presl. Yellowstone Park, August.
Claytonia linearis, Hook. Clark's Fork, August.
Spraguea umbellata, Torr. Yellowstone Park, August.
Dryas octopetala, L. Teton Mountains, July.
Ivesia Gordoni, Gray, Teton Mountains, July.
Saxifraga oppositifolia, L. Teton Mountains, July.
Tellima parviflora, Hook. Teton Canon, Wyo. (?), July.
Heuchera cylindrica, Dougl. Yellowstone Park, August.
Ribes bracteosum, Dougl. Teton Canon, Wyo. (?) August.
Sedum rhodiola, DC. Teton Mountains, July.
Linnaea borealis, Gronov. Yellowstone Park, August.
Townsendia scapigera Eaton, Teton Mountains, July (?).
Solidago serotina, Ait. as *S. gigantea*, Ait. Yellowstone Park, August (?).
Rudbeckia occidentalis, Nutt. Teton Mountains, July.
Crepis Andersoni, Gray, Yellowstone Park, August.
Crepis nana, Richards, Teton Mountains, July.

Vaccinum ovafolium, Smith. Upper Teton Canon, July.
Ledum glandulosum, Nutt. Shoshone Lake, September.
Pentstemon Menziesii, Hook. Teton Mountains, August (?).
Mimulus nanus, H. & A. as *Eunanus Fremonti*, Gray, Crater Hills, August.
Synthyris alpina, Gray, Teton Mountains, July.
Orthocarpus Tolmiei, H. & A. Fort Bridger, by Dr. Leidy.
Lycopus Virginicus, L. Yellowstone Park, August.
Hydrophyllum capitatum alpinum, Wats. Teton Mountains, July.
Nemophila parviflora, Dougl. Yellowstone, July.
Phlox canescens, T. & G. Teton Mountains, July.
Gilia intertexta, Steud. Teton Mountains, August.
Polemonium foliossimum, Gray. Yellowstone Lake, August.
Kochia prostrata, Shrad. Fort Bridger, by Dr. Leidy.
Erigonum salsuginosum, Hook. Fort Bridger, by Dr. Leidy.
Salix reticulata, L. Teton Mountains, July.
Pinus contorta Dougl. Yellowstone Park, August.
Abies Subalpina, Eng. as *A. grandis*, Trail River Mountains, August.
Lemma triscula, L. Yellowstone Park, August
Zannichellia palustris, L. Yellowstone Lake, 1871.
Goodyera Menziesii, Lindl. Teton Mountains, September.
Corallorhiza mutiflora, Nutt. Shoshone Lake, September.
Lloydia serotina, Reich. Teton Mountains and Clark's Fork, July.
Carex rigida, Good. Red Mountain, September.
Carex alpina, Swartz. Uinta Mountains, by Dr. Leidy.
Calamagrostis sylvatica, DC. Teton Mountains, August.
Spartina gracilis, Trin. Yellowstone Park, August.
Pellæa Breweri, Eaton Teton range, August.
Pellæa densa, Hook. Jackson's Lake, September.
Aspidium Lonchitis, Swartz. Teton Mountains, July.
Aspidium spinulosum, Swartz. Teton Mountains, September.

FROM PARRY'S REPORT ON THE BOTANY OF JONES'S EXPEDITION IN NORTHWESTERN WYOMING, 1873.

Aquilegia flavescens, Wats. Yellowstone Park, August.
Aquilegia Jonesii, Parry, Owl Creek range, July.
Delphinium Menziesii, DC. Fort Bridger, June.

Ranunculus occidentalis, Nutt. Little Sandy, June.
Myosurus minimus, L. Snake River, September.
Thalictrum alpinum, L. Wind River range. July.
Stanleya tomentosa, Parry, Owl Creek, July.
Draba ventosa, Snake River Pass, September.
Arabis canescens, Nutt. Sweetwater, June.
Lesquerella (*Vesicaria*) alpina, Nutt. Green River, June.
Capsella divaricata, Walp. Little Sandy, June.
Nasturtium curvisiliqua lyratum, Wats. Yellowstone, August.
Subularia aquatica, Yellowstone Lake, August.
Arenaria Franklinii, Dougl. Wind River, July.
Arenaria pungens, Nutt. Stinkingwater, July.
Arenaria stricta, Wats as *A. Rossii*, R. Br. Owl Creek range, July.
Arenaria arctica, Stev. Owl Creek, July.
Lychnis Drummondii, Wats. Owl Creek, July.
Lychnis Kingii, Wats. as *L. Ajanensis*, Regel. Owl Creek range, July.
Spraguea umbellata, Torr. Stinkingwater, August.
Calyptridium roseum, Wats. Green River, June.
Rhamnus alnifolia, L'Her. Stinkingwater, August.
Lupinus minimus, Dougl. Stinkingwater, August.
Trifolium andinum, Nutt. Ham's Fork, June.
Astragalus ventorum, Gray, Wind River, July.
Astragalus triphyllus, Pursh, Owl Creek, July.
Astragalus simplicifolius, Gray, Green River, June.
Astragalus glabriusculus, Gray, Wind River, July.
Astragalus lotiflorus, Hook. Wind River, July.
Astragalus Geyeri, Gray, Green River, June.
Astragalus flavus, Nutt. Green River, June.
Astragalus pubentissimus, Nutt. Green River, June.
Astragalus glareosus, Dougl. Green River, June.
Astragalus microcystis, Gray, Stinkingwater, July.
Oxytropis campestris, L. Owl Creek, July.
Oxytropis viscida, Nutt. Wind River, July.
Oxytropis lagopus, Nutt. Pacific Springs, June.
Spirea cæspitosa, Nutt. Owl Creek range, July.
Ivesia Gordoni, Gray, Stinkingwater, July.
Heuchera cylindrica Dougl. Stinkingwater, July.
Saxifraga debilis, Engelm. Owl Creek range, July.

Ribes viscosisimum, Pursh, Yellowstone, August.
Ribes bracteosum, Dougl. Wind River, July.
Œnothera andina, Nutt. Green River, June.
Œnothera scapoidea, Nutt. Green River, June.
Zaushneria Californica, Pressl. Stinkingwater, July.
Peucedanum leiocarpum, Nutt. Yellowstone, August.
Cymopterus Fendleri, Gray. Green River, June.
Lonicera cærulea, L. Yellowstone, August.
Kellogia galeoides, Torr. Stinkingwater, August.
Erigeron coccinnus, Gray, Green River, June.
Townsendia spathulata, Nutt. Wind River, July.
Townsendia Watsoni, Gray, as *T. strigosa*, Nutt. Wind River, July.
Townsendia Parryi, Eaton. Wind River, July.
Townsendia condensata, Parry, Washakie's Needles, July.
Bahia (*Schkuhria*) integrifolia, Parry, Wind River Mountains, July.
Rudbeckia occidentalis, Nutt. Snake River, September.
Arnica Parryi, Gray, Yellowstone, August.
Aplopappus suffruticosus, Gray, Yellowstone, August.
Aplopappus multicaulis, Gray, Wind River, July.
Balsamorrhiza Hookeri, Nutt. Pacific Springs, June.
Antennaria dimorpha, Nutt. Green River, June.
Antennaria luzuloides, T. & G. Stinkingwater, July.
Tanacetum Nuttallii, T. & G. Wind River, July.
Artemisia pedatifida, Nutt. Green River, June.
Artemisia spinescens, Eaton, Green River, June.
Artemisia discolor incompta, Gray, Owl Creek, July.
Troximon glaucum parviflorum, Gray, Green River, June.
Stephanomeria paniculata, Nutt. Stinkingwater, July.
Crepis occidentalis, Nutt. Wind River, July.
Laurentia (Porterella) carnosula, Benth. Yellowstone, August.
Ledum glandulosum, Nutt. Yellowstone, August.
Gaultheria Myrsinites, Hook, Yellowstone, August.
Pyrola picta, Smith, as *P. dentata*, Hook. Yellowstone, August.
Androsace Chamæjasme, L. Owl Creek, July.
Douglasia montana, Gray, Owl Creek Mountains, July.
Pentstemon Menziesii, Hook, Stinkingwater, August.
Mimulus nanus. H. & A. as *Eunanus Fremontii*, Gray, Yellowstone, August.

Castilleia breviflora, Gray, Stinkingwater, July.
Gilia pungens, Benth. Green River, June.
Gilia iberidifolia, Benth. Green River September.
Asclepias brachystephana, Torr. Green River, June.
Polygonum imbricatum, Nutt. Stinkingwater, July.
Oxytheca dendroidea, Nutt. Big Sandy, June.
Atriplex endolepis, Watson, Stinkingwater, July.
Grayia polygaloides, H. & A. Green River, June.
Carex aquatilis, Wahl. Yellowstone, August.
Carex rigida, Good. Yellowstone, August.
Carex Hoodii, Boot. Wind River, July.
Carex tenuirostris, Olney, Yellowstone, July.
Isoetes Bolanderi, Engelm. Yellowstone, August.

FROM GRAY'S REPORT ON THE PLANTS OF THE JENNEY SURVEY OF THE BLACK HILLS, 1875.*

Clematis alpina occidentalis, Gray.
Thalictrum dioicum, L.
Aconitum Fischeri, Reichenb.
Lesquerella (*Vesicaria*) alpina, Nutt.
Viola delphinifolia, Nutt.
Helianthemum Canadense, Michx.
Polygala alba, Nutt.
Psoralea esculenta, Pursh.
Astragalus gracilis, Nutt.
Astragalus simplicifolius, Gray.
Lathyrus ochroleucus, Hook.
Sophora sericea, Nutt.
Geum rivale, L.
Rubus triflorus, Richard.
Œnothera pumila, L.
Thaspium trifoliatum, Gray.
Cymopterus glomeratus, Nutt.

Cornus Canadensis, L.
Aster falcatus, Lindl.
Solidago speciosa angustata, T. & G.
Echinacea angustifolia, DC.
Helianthus strumosus, L.
Helianthus petiolaris, Nutt.
Pentstemon grandiflorus, Nutt.
Pentstemon albidus, Nutt.
Verbena bipinnatifida, Nutt.
Lophanthus anisatus, Benth.
Mertensia oblongifolia (Nutt.) Don.
Eriogonum multiceps, Nees.
Lilium Philadelphicum, L.
Prosartes lanuginosa, Don.
Pellaea atropurpurea, Link.
Onoclea sensibilis, L.

* Although the Black Hills are largely in Dakota, yet as the route of the party to and from them lay in this state, and as the flora of the Hills may reasonably be supposed to be approximately the same in both states, these names are included here.

SUMMARY.

In the foregoing lists there have been enumerated from the material in this herbarium 1118 species and varieties of *Phanerogams* (*Spermatophytes*), representing 393 genera. Omitting duplicates from the lists of plants reported by others there are enumerated 177 more, making a total of 1295 thus far reported from this state. This number, undoubtedly, does not do justice to all the work that has been done in the state and falls far short of the number that may be expected when it shall have been thoroughly worked. The northeast and southwest floras are quite distinct from each other and from those portions of the state which have been the most carefully examined. These are yet to be secured.

By way of comparison it may be stated that the following are among the best worked states and the number of species and varieties of *Phanerogams* reported are for Nebraska about 1460, and for West Virginia 1309.

Concerning the *Cryptogams* it may be said that they represent largely an unexplored field. Only 65 species are enumerated in this list, making the total number from this collection 1176 and for the state according to this list 1360. This does not include the Mosses and Lichens of the Hayden report, and possibly other collections may have been omitted.

Duplicates of a large part of the plants enumerated from this collection will be found in the herbaria of Harvard University, Columbia University, Shaw Botanic Garden, National Herbarium, the Vanderbilt collections at Biltmore, N. C., University of Minnesota, Cornell University, and Prof. E. L. Greene's Herbarium.

INDEX TO GENERA.

Abronia	168
Acer	91
Achillea	137
Acerates	150
Aconitum	79
Actæa	79
Æcidium	202
Agrimonia	105
Agropyrum	196
Agrostemma	86
Agrostis	191
Alisma	185
Allium	183
Allocarya	150
Alnus	178
Alopecurus	191
Amarantus	169
Ambrosia	132
Amelanchier	106
Amorpha	94
Ampelopsis	91
Anaphalis	131
Andropogon	189
Androsace	140
Anemone	75
Angelica	117
Antennaria	131
Anthemis	137
Aphyllon	165
Aplopappus	121
Apocynum	150
Aquilegia	78
Arabis	81, 204
Aralia	117
Arceuthobium	176
Arctostaphylos	147
Arenaria	87
Argemone	80
Aristida	190
Arnica	139
Artemisia	137
Asclepias	150
Asplenium	199
Aster	125
Astragalus	95, 205
Atriplex	171
Aulacomnium	200
Bahia	135
Balsamorhiza	133
Barbarea	80
Batrachospermum	202
Barbula	200
Beckmannia	193
Berberis	80
Betula	178
Bidens	134
Bigelovia	122
Bouteloua	193
Brachythecium	201
Brassica	83
Brickellia	119
Bromus	196
Brunella	167
Bryanthus	147
Bryum	201
Buchloe	193
Bupleurum	115
Cactus	114
Calamagrostis	192
Calandrinia	88
Callitriche	110
Calochortus	183
Caltha	78
Calypso	181
Campanula	147
Capsella	84
Cardamine	81
Carex	187
Carum	116
Castilleia	163
Catabrosa	194
Ceanothus	91
Cenchrus	190
Cerastium	86
Ceratodon	200
Cercocarpus	101
Cereus	115
Chænactis	134
Chamærhodos	105
Chara	202
Cheilanthes	199
Chenopodium	170

—16

Chimaphila	148	Erysiphe	203
Chrysopsis	120	Erythronium	183
Cicuta	116	Euphorbia	177
Clarkia	111	Eurotia	171
Claydonia	203	Evernia	203
Claytonia	88	Evolvulus	159
Clematis	75	Festuca	196
Cleome	84	Filago	131
Climacium	201	Fragaria	102
Cnicus	143	Franseria	132
Coldenia	156	Frasera	152
Collinsia	162	Fraxinus	150
Comandra	176	Fritillaria	183
Convolvulus	159	Gaillardia	136
Cordylanthus	164	Galium	119
Cornus	118	Gaura	113
Corydalis	80	Gayophytum	111
Crataegus	106	Gentiana	151
Crepis	143	Geranium	90
Croton	177	Geum	101
Cryptogramme	199	Gilia	153
Cuscuta	160	Glaux	150
Cycloloma	170	Glyceria	195
Cymopterus	116	Glycerrhiza	99
Cystopteris	200	Gnaphalium	132
Dalea	95	Grindelia	120
Danthonia	193	Gutierrezia	120
Delphinium	79	Gymnolomia	133
Deschampsia	192	Habenaria	181
Desmatodon	200	Harbouria	115
Disporum	184	Hedeoma	166
Distichlis	194	Hedysarum	99
Dodecatheon	149	Helenium	136
Draba	82	Helianthus	133
Dracocephalum	167	Heracleum	117
Dysodia	136	Heuchera	107
Eatonia	194	Hieracium	144
Echinocactus	114	Hierochloa	190
Echinospermum	156	Hippuris	110
Eleagnus	176	Hordeum	197
Eleocharis	186	Humulus	177
Ellisia	155	Hydrophyllum	155
Elodia	181	Hymenopappus	135
Elymus	197	Hypericum	89
Epilobium	110	Hypnum	201
Equisetum	199	Ipomea	159
Eragrostis	194	Iris	182
Erigeron	128	Iva	132
Eriogonum	172	Jamesia	108
Eriophorum	187	Juncus	184
Eriophyllum	135	Juniperus	198
Erysimum	83	Kalmia	148

Kœleria	194	Orthocarpus	164
Krynitzkia	157	Oryzopsis	190
Kuhnia	119	Osmorrhiza	116
Lactuca	146	Oxalis	90
Lathyrus	100	Oxybaphus	168
Lepachys	133	Osyris	175
Lepidium	84	Oxytropis	98
Leptobryum	201	Pachystima	91
Lesquerella	82	Panicum	189
Leucocrinum	182	Parietaria	178
Lewisia	89	Parmelia	203
Liatris	120	Parnassia	107
Ligusticum	116	Paronychia	169
Linum	90	Pastinaca	117
Lippia	165	Pedicularis	165
Listera	181	Peltigera	203
Lithospermum	159	Pentstemon	161
Lonicera	118	Petalostemon	95
Lophanthus	167	Petasites	139
Lupinus	92	Peucedanum	117
Luzula	185	Phacelia	155
Lycopus	160	Phalaris	190
Lygodesmia	146	Philonotis	200
Madia	134	Phleum	191
Malvastrum	89	Phlox	152
Marchantia	202	Phragmidium	203
Medicago	93	Phragmites	194
Melampsora	203	Physalis	160
Melica	194	Physaria	82
Meliilotus	93	Physocarpus	101
Mentha	166	Physostegia	167
Mentzelia	113	Picea	197
Mertensia	158	Pinus	197
Mimulus	162	Plantago	167
Mitella	107	Pleurogyne	152
Mnium	201	Poa	194
Monarda	166	Polanisia	85
Moneses	148	Polemonium	154
Monolepis	170	Polygonum	174
Monotropa	148	Polytrichum	201
Muhlenbergia	190	Populus	180
Munroa	193	Portulaca	88
Musenium	115	Potamogeton	186
Myosotis	159	Potentilla	102
Myriophyllum	110	Primula	149
Nasturtium	80	Prunus	100
Negundo	91	Pseudoleskia	201
Nuphar	80	Pseudotsuga	198
Œnothera	112	Psoralea	93
Onosmodium	159	Pteris	200
Opuntia	115	Pterospora	148
Oreoxis	117	Purshia	101

Pyrola	148	Sporobolus	191
Pyrus	105	Stachys	167
Quercus	178	Stanleya	83
Ramularia	203	Steironema	150
Ranunculus	76	Stellaria	87
Rhus	91	Stephanomeria	146
Ribes	108	Stipa	190
Rinodina	203	Streptopus	182
Rosa	105	Sueda	172
Rubus	101	Swertia	152
Rudbeckia	133	Symphoricarpos	118
Rumex	175	Synthyris	163
Sagittaria	186	Tanacetum	137
Salicornia	172	Taraxacum	145
Salix	178	Tellima	107
Salsola	172	Tetradymia	142
Salvia	166	Thalictrum	75
Sambucus	118	Thelesperma	134
Sanicula	115	Thelypodium	81, 204
Saponaria	86	Thermopsis	92
Sarcobatus	172	Thlaspi	84
Saxifraga	106	Timmia	201
Schedonardus	193	Tissa	88
Scirpus	186	Townsendia	125
Scrophularia	160	Tradescantia	184
Scutellaria	167	Trifolium	94
Sedum	110	Triglochin	186
Selaginella	200	Trisetum	192
Selinum	117	Trollius	78
Senecio	140	Troximon	145
Setaria	190	Typha	185
Shepherdia	176	Uromyces	203
Sibbaldia	104	Urtica	177
Sidalcea	89	Vaccinum	147
Silene	86	Valeriana	119
Sisymbrium	182	Verbena	166
Sium	116	Veronica	163
Smelowskia	83	Viburnum	118
Smilacina	182	Vicia	99
Solanum	160	Viola	85
Solidago	124	Vitis	91
Sonchus	156	Woodsia	200
Spartina	103	Wyethia	133
Specularia	147	Yucca	182
Sphæralcea	90	Zizia	116
Spiræa	100	Zygadenus	184
Spiranthes	181		

www.ingramcontent.com/pod-product-compliance
Lightning Source LLC
Chambersburg PA
CBHW020253170426
43202CB00008B/351